MIDLAND TIMES

CONTENTS

Introduction	3
Forres Remembered	4–13
Motive Power at Bolton shed during the latter years	14–23
LMS-built Locomotive Naming Strategy Part I – Black Fives and Patriots	24–51
The Caledonian Railway Class 294 'Jumbo' 0-6-0s	52–59
The Midland 'Singles'	60–65
The Railways of Leicester	66–69
Hest Bank	70–77
OUT NOW/COMING SOON:	
The R.C.Riley Colour Collection	78
The Platform End	79–80

13TH SEPTEMBER 1958 • Stanier 'Coronation' Class 8P 4-6-2 No. 46252 CITY OF LEICESTER at Hest Bank. There are more images taken at Hest Bank in this issue, starting on page 70.
PHOTO: NEVILLE STEAD COLLECTION © THE TRANSPORT TREASURY

© Images and design: The Transport Treasury 2025. Design and Text: Peter Sikes
ISBN: 978-1-917776-26-4
First published in 2025 by Transport Treasury Publishing Ltd., 16 Highworth Close, High Wycombe HP13 7PJ
With thanks to Dennis Troughton

The copyright holders hereby give notice that all rights to this work are reserved.
Aside from brief passages for the purpose of review, no part of this work may be reproduced, copied by electronic or other means, or otherwise stored in any information storage and retrieval system without written permission from the Publisher. This includes the illustrations herein which shall remain the copyright of the copyright holder.

Copies of many of the images in MIDLAND TIMES are available for purchase/download.
In addition, the Transport Treasury Archive contains tens of thousands of other UK, Irish and some European railway photographs.

www.ttpublishing.co.uk or for editorial issues and contributions email MidlandTimes1884@gmail.com

Printed in England by Short Run Press Limited, Exeter.

INTRODUCTION

We start the ninth issue of *Midland Times* with an article titled, Forres Remembered, an interview and reminiscence by Ian Lamb with Alistair Barrie about railway life at Forres, an important Highland junction. It includes photos and details of operations, signal boxes, staff and locomotives from the 1930s to the 1960s.

Bolton shed is the next subject in our motive power series from the latter days of BR(M) steam operation, examining the locomotives and final days of the depot up to its closure in 1968.

We then embark on an in-depth look at LMS-built Locomotive Naming Strategy. Intending this to be a series, part one looks at Black Fives and Patriots. Philip Hellawell explores the conventions, reasoning and history behind the naming practices for key LMS locomotive classes, focusing on regional and commemorative names.

David Anderson then takes a look at the history of one of the most numerous class of engines on Scottish railways, the Caledonian Class 294 'Jumbo' 0-6-0s. It is a detailed technical and historical account of Dugald Drummond's long-lived Scottish freight locomotives, known for their durability and simplicity.

Midland 'Singles' is our next offering, looking at the Midland Railway's elegant and iconic 4-2-2 locomotives. David Cullen describes the single-driver locomotives – the '*Spinners*' – with extensive mechanical detail and context on their 19th-century development.

The Railways of Leicester is a short historical overview of Leicester's railway stations and their evolution. There are a few images from the Midland Railway lines at West Bridge and London Road which will hopefully whet your appetite for a deeper look into the city's railways – and I unashamedly reference my new book '*The Railways of Leicester*' which is available to purchase from Transport Treasury Publishing.

The final offering for this issue is a look at Hest Bank, as referenced in the photo on page 2. It features a history of Hest Bank station, its role in the West Coast Main Line and its eventual closure. It includes an illustration of the station's layout and concludes with a photographic essay around the coastal station featuring classic LMS motive power.

As usual, my thanks go to the readers who took time to write in with their observations and these are reproduced in 'The Platform End'. It is always good to hear from you especially with suggestions of locations you would like to see featured in *Midland Times* and also pointing out facts and additional information that have not been covered in the articles produced, your input is much appreciated.

To make sure you never miss an issue of *Midland Times* why not sign up to our subscription service? For details visit **www.ttpublishing.co.uk**, email **admin@ttpublishing.co.uk** or call us on **01494 708939**.

PETER SIKES, EDITOR, MIDLAND TIMES
email: midlandtimes1884@gmail.com

FRONT COVER/LEFT: Stanier Class 6P 'Jubilee' 4-6-0 No. 45738 SAMSON parked up at the south end of Carlisle Citadel station. Built at Crewe in 1936 it was allocated to Bushbury, Crewe North, Edge Hill and Camden before ending its working life at Carlisle Upperby (12B) and finally Carlisle Kingmoor (12A), from where it was withdrawn on 28th December 1963.
PHOTO: R. C. RILEY
© TRANSPORT TREASURY

MIDLAND TIMES • ISSUE 9

FORRES REMEMBERED
AN INTERVIEW WITH ALISTAIR BARRIE
by Ian Lamb

27TH AUGUST 1965
A group of excited lads greet the arrival of Highland Railway Jones Goods No. 103 at Forres. This was one of the Highland Railway Centenary round trips from Inverness (there was one in the morning and another in the afternoon). No. 103 is hauling the two restored Caledonian Railway coaches, numbered 7396 and 3339.
PHOTO: NORRIS FORREST © THE TRANSPORT TREASURY

Fate does play a strange hand at times, and this was no different with my long association on railway matters with Alistair. I was a child in the western suburbs of Edinburgh, yet have spent most of my life in the Highlands. The opposite happened with Alistair who grew up in Forres, yet has been mainly domiciled almost a "stone's throw" from my place of youth.

Our common denominator was The DAVA Project, established as a Millennium event in 2000 to encourage local young people in Strathspey to take an interest in the railways that once ran through the valley, and support the heritage railway that was trying hard to ensure the past would not be forgotten by running preserved trains between Aviemore and Grantown-on-Spey.

Alistair was very much a keen railway enthusiast during the early sixties, not least of all because Forres was a very busy railway junction with regular train services from Inverness, Aberdeen and Perth passing through. These services served the local community and surrounding areas, also the county of Moray, where numerous servicemen and women returned to the Royal Air Force base at Kinloss or the Royal Naval Air Station at Lossiemouth.

There were three signal cabins controlling the junction – Forres West, Forres South and Forres East, which opened between 5am and 6am; closing between 10pm and 11pm.

The station was worked on a shift basis with an early allocation of foreman and porter, followed by a back duty of foreman and porter, plus a junior porter on day shift. Forres booking office was also an early shift, back duty and for additional staff on day shift. Responsibility for overseeing not only the station but also the goods shed and delivery staff, was the Station Master.

Duncan McKay (a Forres man), was believed to be the last Station Master of this once important Forres Junction. Prior to his appointment at Forres, he was relief Station Master for Inverness, and one of the senior operational railwaymen in the North of Scotland.

The first station was opened in Forres on 25th March 1858 when the line from Nairn was extended through the town eastwards to Elgin. Forres original station was closed and replaced in a triangular fashion with the opening of the line south over the Dava and Drumochter summits to Perth in 1863.

With closure of the engine shed in 1959 and transfer of the remaining steam locomotives to other sheds – or awaiting the scrap man's torch – Forres (60E) received a new 350 BHP English

JUNE 1952 • Forres station was located on a triangle; an arrangement that allowed a connection to the original main line to Aviemore from either the Inverness or Elgin direction. Here, Standard Class 5MT No. 73007 waits at the platform with a local service for Aviemore. PHOTO: JIM FLINT/JIM HARBART © THE TRANSPORT TREASURY

ABOVE: 16TH JUNE 1962 • A view of the old station building (now in the goods yard) looking towards Elgin showing the east signal cabin and crossing. PHOTO: LESLIE FREEMAN © THE TRANSPORT TREASURY

BELOW: 23RD MAY 1950 • Forres engine shed.
PHOTO: NEVILLE STEAD COLLECTION © THE TRANSPORT TREASURY

Electric diesel shunter number D3735. 'The Diesel' (as it became known in railway circles) was used on the daily trips (Monday/Saturday) to Burghead, departing Forres Yard around 8am and returning from Burghead after midday. The diesel shunter was also used on the Mondays only trip to Dallas Dhu Distillery, leaving Forres Yard between midday and 1pm following departure of the Perth Goods, returning from Dallas Dhu between 2pm and 3pm.

D3735 was moved from Forres to Inverness in the early 1960s, and replaced by one of Barclay 0-4-0 diesel shunters (later class 06) allocated to Inverness. This Barclay loco was used on the Burghead trips but the local P-way Inspector Mr. Bryden complained about the damage to the track. Eventually D3896 was transferred from Aviemore to replace the Barclay shunter. Closure had been under consideration in 1956, but was deferred for the highly novel reason "that skilled older men would not care to move, and thus be lost to the railway service".

From 1st October to 31st March (Winter months) steam engines, and later diesel locomotives, were required to be snowplough-fitted for travelling over 'The Hill' (Drumochter), Dava and Slochd summits.

The Swindon built Cross-Country three-car Diesel Multiple Units were introduced on the Inverness to Aberdeen services via Mulben and Keith, being allowed two and a half hours for the journey. Initially this service was four return trips daily (Monday/Saturday, no Sunday service).

There were other Inverness to Aberdeen services – some via the former GNSR coast route – with five coach trains of mainly Stanier stock until replaced by BR Mark 1s, hauled by B1, BR Standard 4MT 2-6-0 and BR Standard 4MT 2-6-4T locomotives until replaced by North British Type 2 diesels D6138 to D6157, locally known as *Mans*, all allocated to 61A Kittybrewster Aberdeen. In some respects, the services in BR days were reversed from that of the original Highland Railway in-so-far as the one-time stranglehold held by Inverness was now that of Aberdeen! (As far as is known, no Inverness drivers were passed for the 'NB' Type 2s).

The principal services were the Up and Down trains over '*The Dava*', Inverness, Forres to Aviemore. The services continued onward to Perth, Edinburgh, Glasgow and destinations in England.

One of the interesting services was the 12.17pm train from Perth to Inverness via '*The Dava*' and Forres, due in Forres around 4.20pm, which stopped at all stations. This train consisted of a BCK from Edinburgh (Waverley), RB as far as Aviemore, BSK and CK from Glasgow (Buchanan Street), BG from Perth, plus a 'Vanfit' on Tuesdays to carry yeast.

During the summer holiday season (especially the 'Glasgow Fair') this would have additional coaches and possibly require two locomotives. Normal motive power could be a '*Hiker*' (Black 5), BR Standard 5, Birmingham or Derby (BRC&W or BR Type 2s, later Class 26 and 24). The crew for this service had taken '*The Royal Highlander*' sleeping car train from Inverness to Perth the previous evening and lodged overnight in the Perth Railway Hostel.

The locomotive for this service had spent the previous day on the Aberfeldy Branch, working into Perth earlier in the day. On Tuesdays a 'Vanfit' was added with yeast for shipment to

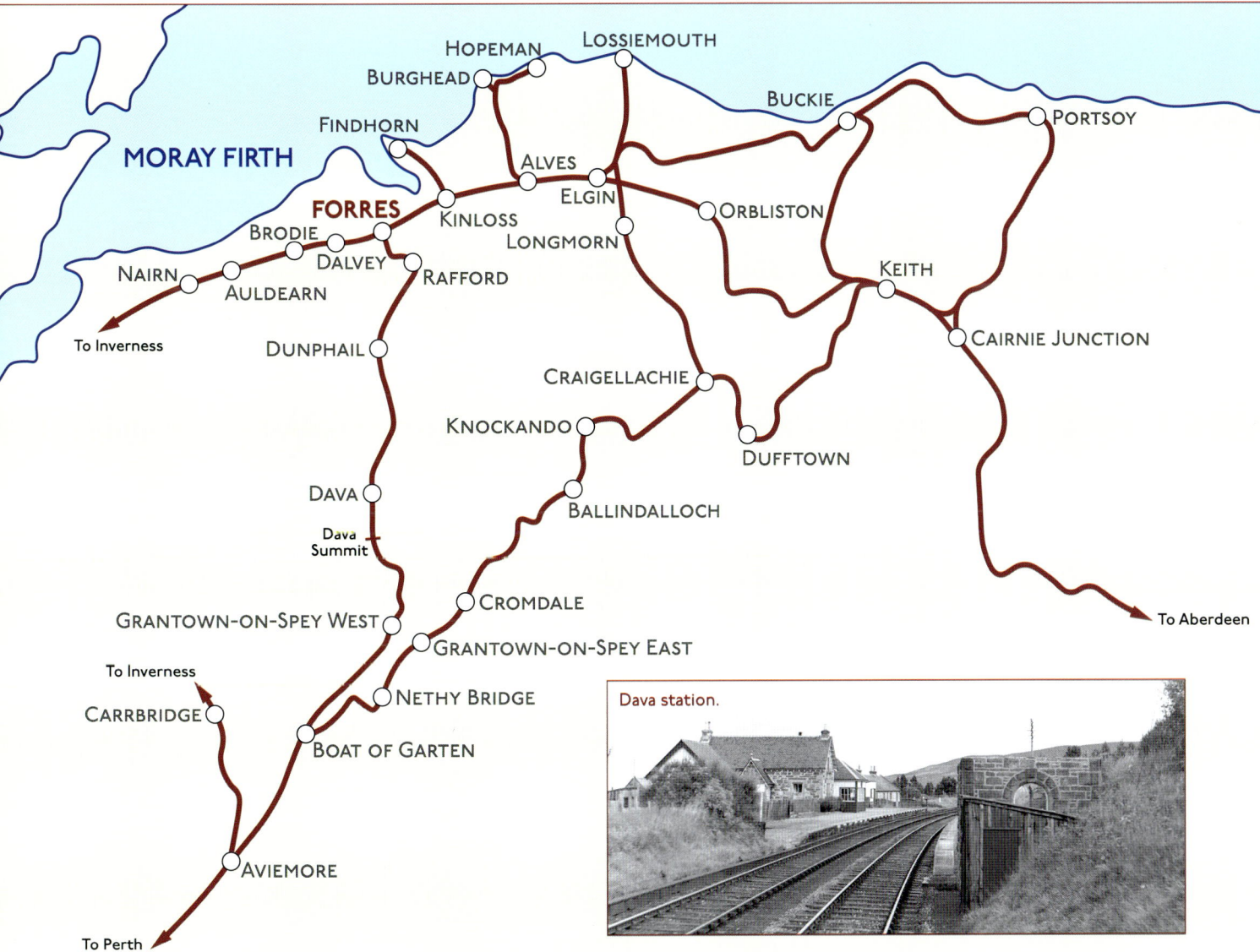

Dava station.

Distilleries in the north. Also, Dallas Dhu, Benroamach and Glenburgie distilleries in and around Forres each received two or three half hundred weight bags of yeast. It was normal for the railway delivery lorry to appear and convey this product to one of the distilleries "in case it was required", then the driver could join the workers and claim a glass of whisky!

Another interesting service was the (circa) 4.40pm Inverness to Aberdeen (via coast), arriving in Forres at 5.26pm which included *The Royal Highlander* Forres portion. This consisted of SLC, CK and BSK coaches bound for London Euston, shunted by the Forres diesel shunter/pilot on to the train in Platform 1, headed by a *'Hiker'* (Black 5) in steam days, or a Birmingham or Derby (BRC&W or BR Type 2s, later class 26 and 24) until the withdrawal of services.

HRH Prince Philip, The Duke of Edinburgh, used this service on many occasions to return to London following private visits to his sons at Gordonstoun. In today's security conscious world, Prince Philip waiting on the platform at Forres with his bodyguard for the train to arrive is well remembered.

On Saturday evenings during the Summer timetable, a Park Royal railbus provided an Aviemore 7.01pm departure, over *'The Dava'* to Forres arriving around 8.10pm, then on to Inverness. Various types of railbuses were used to reduce costs, but were not really successful because of all the other operational staff required. Possibly with the future radio signalling and other initiatives, more of these lines would still be open today. The locals nicknamed them *'Sputniks'* due to their ungainly shape and reference to the Russian space orbit.

During the summer months on Friday evenings there was a York to Inverness car carrier train. Motive power from Perth could be two 'NB' Type 2s (D6100-D6137), with the same locos being used on Saturday mornings to head the (circa) 10am Inverness-Aviemore service (via Forres), joining the main train for onward travel to Perth, Edinburgh and Glasgow. The locos then returned (presumably with a direct Inverness service) before heading south with the Inverness to York car carrier train.

On occasion one of Aviemore's *'Hiker Pugs'* (Fairburn 4MT 2-6-4 tank engines) would be the locomotive on the Aviemore to Forres services – morning mail or paper trains.

Goods trains over *'The Dava'* consisted of the daily Perth-Forres goods, and return Forres-Perth goods. There was also a Class 6 (through Freight) from Milburn Yard, Inverness with a 4.30pm departure via Forres for Aviemore. If the loco was a Derby (BR Type 2, later class 24) the headcode would read 4E30 (class E/6, 4.30pm from Milburn Yard). Traffic from Forres could be wagons with butts of whisky for blending, seed potatoes, timber for pit props, sugar beet for Cupar, empty 16 ton end doors returning to Fife coal fields and empty Briggs tar tanks.

There were also goods trains on the Inverness, Forres, Elgin to Aberdeen and return services, with specials of cattle or sheep, following sales.

On Sundays the *'New Road'* (via Carrbridge) was closed and the two Down services and two Up services used *'The Dava'*. The Down *'Royal Highlander'* left London Euston on Saturday evening, arriving in Forres 9.45am. The Down service from Edinburgh and Glasgow arrived in Forres around 12.59pm.

7TH SEPTEMBER 1958 • Ex-Caledonian McIntosh Class 439 (2P) 0-4-4T No. 55178 rests at the side of the Permanent Way building PHOTO: DAVID ANDERSON © THE TRANSPORT TREASURY

Up trains were the 3.30pm to Edinburgh and Glasgow which included the Highland Railway POS (Travelling Post Office Sorting Van), being replaced by a new POS based on the Mark 1 coach. This train left Forres around 4.03 pm. The 'Royal Highlander' left Inverness at 5pm and due to arrive in Forres at 5.30pm, departing at 5.37pm. This schedule was very difficult to maintain as the train could be loaded to 540 tons or more with a maximum of 17 coaches on occasion, especially around the time of the 'Glorious 12th' shooting season. Also included in that consist was an RU for Perth and a BG for Kilburn (Euston being redeveloped)

As a result of this heavy load three Type 2 locomotives (BRC&W or BR, Birminghams or Derbys – some crews preferred one of each, plus a third) would be provided for the haul to Dava Summit (1,052 feet above sea level) and the onward haul from Aviemore on 'The Hill' or as it is known Drumochter Summit (1,484 feet above sea level).

On Saturday 19th February 1955, Buckie Thistle FC were hosts to Heart of Midlothian FC in the Scottish Cup. The match resulted in a 6-0 win for Hearts. The next day Hearts headed back to Edinburgh from Forres railway station on the 3pm train. Parties of boys were allowed on the team bus by Hearts trainer, John Harvey, to obtain players' autographs, while others were on the station platform throwing snowballs! At the same time, Alistair's father was talking with the Hearts manager, the great Tommy Walker.

Saturday 23rd May 1964 – Deltic D9004 was named *Queen's Own Highlander* at Inverness, prior to working the Up 'Royal Highlander' to Perth, then light engine home to Haymarket. Duncan McKay was acting Station Master at Inverness at that time, on seeing Alistair he returned to his office to get a platform pass for the lad to see D9004 being named at around 4.00pm. Harold Wilson, the then leader of the Labour Party, the Honourable member for Huyton near Liverpool (he was sponsored by the NUR), was on the platform at Inverness and it is believed that he travelled south on the 'Royal Highlander'. Alistair travelled to Aviemore on that train then home to Forres over 'The Dava' by Park Royal railbus (summer Saturday service).

LEFT: 16TH JUNE 1962 • The junction looking towards Inverness with Forres West signal cabin at its throat. A Type 1 English Electric diesel (known locally as '1000s') shunts in Forres goods yard. To the left is the original Highland Railway main line to Aviemore (via Dava). A small goods train stands in the South siding. In the centre are the tracks between Inverness and Aberdeen, with a preserved Caledonian Railway coach resting at platform 3. The engine shed was located just to the right of this photograph.
PHOTO: LESLIE FREEMAN © THE TRANSPORT TREASURY

1930s • Former Highland Railway Jones 'Loch' class 4-4-0 No. 14385 LOCH TAY at Forres shed. Note the turntable pit behind the tender, and the engine shed beyond.
Photo: Sandy Murdoch © The Transport Treasury

1936 • Ex-Highland Railway Drummond Class U 4-4-0 No. 14422 Ben Achaoruinn at Forres.
Photo: Neville Stead Collection © The Transport Treasury

An undated shot of ex-Caledonian Superheated Class 113 (3P) 4-4-0 No. 14472 in LMS days with an eastbound train from platform 3.
Photo: Jim Flint/Jim Harbart © The Transport Treasury

Another undated shot, but probably taken in 1949/50, of Pickersgill Class 72 (3P) 4-4-0 No. 54481 departing from platform 2 at Forres station with a train bound for Inverness, platform 1 curves away to the right with the tracks for Aviemore (via Dava).
Photo: Neville Stead Collection © The Transport Treasury

21ST APRIL 1962 • Highland Railway Jones Goods 4-6-0 No. 103 on an SLS/BLS Scottish Rambler tour at Forres being passed by a Cross Country set on the 8.30a.m. Inverness-Aberdeen service.
PHOTO: W. A. C. SMITH © THE TRANSPORT TREASURY

16TH JUNE 1962 • Stanier Black Five 4-6-0 No. 44978 arrives into the north platform 3 at Forres at the head of the Railway Correspondence and Travel Society/Stephenson Locomotive Society Joint Scottish Tour. PHOTO: LESLIE FREEMAN © THE TRANSPORT TREASURY

21ST MAY 1960 • Ex-Caledonian Railway Pickersgill Class 72 (3P) 4-4-0 Nos. 54485 and 55486 pause at Forres while working the Up Highland Mail after the filming of the Down Mail from Perth to Inverness via Slochd. PHOTO: JOHN MCCANN - ONLINE TRANSPORT ARCHIVE/TRANSPORT LIBRARY

MOTIVE POWER AT BOLTON SHED DURING THE LATTER YEARS
by David Young
Courtesy of Chris Tasker, Manchester Locomotive Society

Bolton Shed, known as 'Crescent Road', but with the shed entrance located off Back Crescent Road, was on the west side of the Clifton Junction–Bolton (Trinity Street) main line. The shed was opened by the LYR in 1874, replacing an earlier building which had been erected by the Manchester and Bolton Railway in 1840, on the opposite side of the main line, and known as 'Burnden'. In 1935, Bolton shed, now coded 26C, was extensively modernised by the LMS, and was re-roofed and equipped with a concrete-framed ash plant, followed in 1936 by a new mechanical coaling plant. In 1955, a 60 foot vacuum-operated turntable replaced the 50 foot electric turntable which had been installed by the LMS in 1936.

Bolton had two sub-sheds before 1923, at Horwich and Chorley, but these closed shortly after that date. Primarily a local passenger and freight-oriented shed, 26C had housed a considerable allocation of ex-LYR 0-6-0s and 0-6-0STs, together with 0-8-0s for the longer-distance freights and the well-known 2-4-2 'Radial' tanks and Hughes 'Baltic' 4-6-4 tanks for the suburban and shorter-distance passenger work. Mention should also be made of the 0-4-0T Railmotors, the last of which (10617) survived until 1948, after working the Horwich-Blackrod branch.

Many of the ex-LYR locos were long-lived and, in LMS days, the main changes at 26C were the introduction of the Stanier 2-6-4Ts and the replacement of the ex-LYR 0-8-0s by Fowler 'Austin Seven' 0-8-0s. In addition, 26C was the main 'running-in' base for locos outshopped from Horwich Works after overhaul there or having been newly constructed there and this practice continued after 1948.

In 1945, 26C housed the following 40 locos:		
Locomotive	Numbers	Total
LMS 4MT Stanier 2-6-4T	2633/51-7	Total 8
LMS 4MT Fairburn 2-6-4T	2678	Total 1
LMS 7F Fowler 0-8-0	9573, 9641/64	Total 3
LYR 'Railmotor' 0-4-0T	10600/17	Total 2
LYR 2P 'Radial' 2-4-2T	10646/60-1/97, 10796, 10807/10/31/59/65	Total 10
LYR 2F 0-6-0ST	11348, 11511/3/9	Total 4
LYR 3F 0-6-0	12157, 12212/36/63, 12348/57/65/83, 12404/46, 12528, 12617	Total 12

LEFT: 18TH MAY 1968 • Stanier Black Five No. 45104 at its home shed of Bolton alongside 8F 2-8-0 No. 48168. Allocated to the shed in April 1965, withdrawn on 29th June 1968. PHOTO: TRANSPORT TREASURY

By Nationalisation, (1st January 1948), the allocation had increased to 45 locos, as follows:		
Locomotive	Numbers	Total
LMS Compound 4P 4-4-0	1103-4/90-1/9	Total 5
LMS Stanier 4P 2-6-4T	2545/65, 2633/52-7	Total 9
LMS Stanier 8F 2-8-0	8437/53/68, 8767	Total 4
LMS Fowler 7F 0-8-0	9542/73, 9641/64	Total 4
LYR 'Railmotor' 0-4-0T	10617	Total 1
LYR 2P 'Radial' 2-4-2T	10642/6/97, 10807/15/31/59	Total 7
LYR 2F 0-6-0ST	11348, 11511/3/9	Total 4
LYR 3F 0-6-0	12157, 12212/36, 12348/50/7/65, 12404/46, 12528, 12609	Total 11

During the 1950s, 26C acquired ex-LMS Fowler 2P 4-4-0s for working stopping trains from Bolton to Liverpool Exchange via Wigan Wallgate and 40586 and 40681 of this class were 'regulars' on this duty. Otherwise, Stanier 2-6-4Ts handled most of the passenger jobs worked from 26C, these taking the locos to Blackpool, Southport, Blackburn, Hellifield, Skipton and Todmorden. The 5.40 pm Manchester Victoria-Hellifield, with nine non-corridors, was timed from Victoria to Bolton non-stop in 16 minutes with 42563, an excellent run and quite typical.

The 'Radial' 2-4-2Ts continued to find work in the Bolton area, with 50731 acting as station pilot at Trinity Street, together with 50647/60, which were push-and-pull fitted for use on the former Horwich railmotor turn in addition to the station pilot duties. However, as late as November 1955, 50855 worked Bolton-Blackburn turns in place of the customary 2-6-4T and, in 1959, 50850 became a regular performer on Bolton-Horwich workings, hauling three non-corridor coaches. In February 1960, 50850 was transferred to Southport for station pilot duties. When withdrawn in late 1961, it was the last member of its class in traffic but, unfortunately, was not preserved.

The ex-LYR 0-6-0STs, of which 51408 was a 'favourite', shunted at Rose Hill, Halliwell and Astley Bridge and worked a daily pick-up goods between Bullfield and Turton, calling at Craddock Lane en route. The 0-6-0 locos of ex-LYR Class 'A' shared these turns and 52348, reputedly the best of its class at 26C, had a regular job in 1957 – the 6.5 pm Bolton-Moston freight, then a parcels train from Ashton Moss sidings. Other members of the

ABOVE: An undated shot of ex-LYR Hughes 'Railmotor' 0-4-0T No. 10617. Put into service in December 1911, No. 10617 was the final survivor of the class, working the 'Horwich Jerk' service from Horwich to Blackrod, which was the last part of the L&Y system which made use of Hughes Railmotors.

9TH AUGUST 1953 • 1896 built ex-LYR Class 1008 2-4-2T No. 50731 at Bolton Shed with classmate No. 50887.
BOTH PHOTOS: NEVILLE STEAD COLLECTION © TRANSPORT TREASURY

class, 52523/45 were still active on local freights in the Bolton area as late as October 1962. Interestingly, 350hp diesel-electric shunter D3779 was allocated to 26C in August 1959 as a possible replacement for the 0-6-0STs but cannot have found favour, as it moved to Newton Heath in December that year, and 26C – recoded 9K in 1963 – remained essentially a steam shed until it closed on 1st July 1968.

26C had a Fowler 0-6-0 dock tank, 47165, on its books from November 1961 to November 1963 and this small loco regularly shunted at Halliwell, replacing 51408.

The Fowler 'Austin Seven' 0-8-0s regularly worked the 'Kearsley Pilot' turn, which involved taking 16 loaded coal wagons up the short single mineral line from Kearsley Sidings to Linnyshaw Moss, where the line joined the NCB colliery network. The wagons came from collieries in the Salford area, being brought to Kearsley by Agecroft engines. By the time these 0-8-0s were superseded (49544 and 49618 were at 26C until January 1960), 26C had acquired replacements in the form of WD 2-8-0s 90102/10/206/67/97/641/725/9.

During the mid-1930s, 26C had four 'Prince of Wales' 4-6-0s (ex-LNWR) for passenger work but, until Stanier Class 5 45290 arrived in February 1962 from Newton Heath, no 4-6-0s were on allocation at 26C (unless an ex-LYR Hughes 'Dreadnought' 4-6-0 was based there in the interim). 45290 was considered to be the best Class 5 at 26C by the late Jim Markland and its arrival was followed by no fewer than 45 others of the class, including 45156 *Ayrshire Yeomanry*, (was this 26C's only named engine?) and was there from December 1962 to May 1963. The now-preserved 44871 and 45110 also spent short periods at 26C, as is well known.

The arrival of 45290 at 26C was a turning point in the shed's allocation, as, from the early 1960s, engines displaced elsewhere by dieselisation or line closures came to 'secondary' sheds such as Bolton and increasingly took over work previously performed by smaller locos. A good example would be Standard Class 2 2-6-2Ts 84013-4/7/9/25-7, which 26C acquired between 1959 and 1964, at various times, and of which 84025-7 had initially worked on the Southern Region, in the Ashford and Ramsgate areas. The locos worked the Horwich-Blackrod and Chorley auto-trains and, in fact, 84025 worked the last Horwich-Chorley 'Motor on 24th September 1965. The locos also worked from Bolton to Radcliffe Central via Darcy Lever and Bradley Fold, a route which ex-LYR 'Railmotor' 10617 had once worked prior to its withdrawal in 1948.

In September 1962, Standard Class 9F 2-10-0s 92015-7 moved from Newton Heath to 26C to work the 8.20 pm Bolton-Maston and late evening Ancoats-Heysham freights, which had been Class 5 4-6-0 jobs. The 9Fs had a short stay at 26C, being returned to Newton Heath in December 1962 and replaced at 26C by Class 5s 45156/224/32, which took over the former 9F duties. Perhaps the Class 5s were capable of handling the duties more economically than the 9Fs?

By October 1964, the WD 2-8-0s at 26C had all been transferred away and replaced by Stanier 8F 2-8-0s, of which 26C (later 9K) operated a large number during the mid- to late-1960s. The now-preserved 48773 was there from September 1964 to July 1968 prior to moving to Rose Grove. The 8Fs, as with the WDs before them, became the mainstay of the freight workings from Bolton across to Yorkshire via the Calder Valley route, sharing the route with the numerous Class 5 4-6-0s at Bolton. In this connection,

15TH SEPTEMBER 1961 • Fowler Class 2F 0-6-0T Dock Tank No. 47165 at Irwell Street Yards, Salford. It was allocated to Bolton from November 1961 until November 1963.
PHOTO: ROBERT ANDERSON/TRANSPORT LIBRARY

Stanier Class 3MT 2-6-2T No. 40130 of Llandudno Junction shed is seen with a Horwich to Manchester Victoria train near Lostock Junction while running-in after overhaul at Horwich Works. PHOTO: J. DAVENPORT/D. A. YOUNG COLLECTION © MANCHESTER LOCOMOTIVE SOCIETY

it is noteworthy that Bolton also acquired Standard Class 5s 73004/13-4/9/26/8/40/8/66/9/70 and 73156 in April 1966 and kept 73069 (regarded as the best) until the shed closed. 73156 still survives and is in service at the GCR at Loughborough. Bolton also acquired Standard Class 2 2-6-0s 78007/12-3/23/8/44/55/62 in late 1966 but their stay was short, to May 1967. They carried out duties which would in earlier years have been given to ex-LYR Class A 0-6-0s. 78013 is recorded as having worked at Halliwell on the last day of steam there.

Bolton engines did sometimes 'get into the limelight', in addition to the instances already quoted. Stanier 2--6-4T 42626, specially cleaned by enthusiasts, handled the final passenger workings on the Horwich branch on 25th September 1965. The last train, 12.05 Horwich - Bolton, left to a 60-detonator salute, the loco carrying a wreath and headboard.

Two months before withdrawal, ex-LYR 0-6-0 52523 worked a Roch Valley Railway Society special train on 28th July 1962 from Tottington Junction to Bacup, then Bury Bolton Street, Bury Knowsley Street, Heywood and Castleton to Rochdale.

When Bolton shed closed, Class 5 44781 was one of the allocation, moving to Carnforth prior to participating in the famous 'Fifteen Guinea Special' on 11th August 1968, with 44871. 44781 was used in a film at Bartlow in Essex afterwards and was intentionally derailed. It had been hoped to save the loco for preservation but the cost of movement back to Carnforth was considered excessive and the loco was cut up on site by King's of Norwich.

Otherwise, much of the work carried out by Bolton engines went unrecorded. During the 1960s, some of the Stanier 2-6-4Ts were replaced by Fairburns of similar capacity and, of the latter, 42207/40 had been Scottish-based and 42249/52 came from the ex-LTSR lines. Several 'Crab' 2-6-0s came during the 1962-4 period, mostly for freight work, but did not stay long before giving way to Class 5s and a few 'Derby Four' 4F 0-6-0s also spent time at Bolton. One of the latter was 44311, which, together with the 'Crabs', was often used to bank freight trains from Bradshawgate up the seven-mile ascent to Walton's Siding, before Sough Tunnel on the Blackburn line.

After Bury shed closed in April 1965, Bolton engines took over some of the duties and, in September of that year, 45258 worked a Bury (Knowsley Street)-Blackpool Illuminations excursion, formed of non-corridor stock and travelling via Bury (Bolton Street), Ramsbottom, Helmshore, Haslingden, Accrington, Blackburn and Preston. In March 1966, 44816 was noted shunting Heap Bridge goods yard, which was close to the local paper-making firm Yates Duxbury.

At this time, Bolton's Class 5s were still working passenger services between Rochdale, Bolton, Wigan and Liverpool, sharing the work with the Fairburn 2-6-4Ts. The last day of steam power on these turns was 16th April 1966 and 42663 and 45411 from Bolton were specially cleaned for the occasion.

On 17th March 1968, two railtours ran from Stockport to Carnforth via Bolton. The first used 4472 *Flying Scotsman* to Bolton, then 45290 was coupled 'inside' to assist up the 1 in 72 to Walton's Siding. The second tour, with 70013 Oliver Cromwell, was similarly assisted by 45110.

On 20th April 1968, 45110 was paired with 44949 (9D) and worked from Stockport via Buxton, Chinley and Romiley to Stalybridge, where 73134 (9H) and 73069 came on to the train, which then continued via Standedge, Copy Pit and Sough to Bolton, where 48773 continued via Bury, Rochdale and Royton Junction to Stockport. On 27th April, the tour was repeated, when 73050 (9H) and 73069 covered the Stalybridge-Bolton section and 48652 continued to Stockport.

On 2nd April 1968, 0-6-0 diesel-mechanical shunters D2515/6 were noted on Bolton shed. These had been withdrawn from Barrow shed the previous September but were still in good condition externally.

Bolton shed closed to steam on 1st July 1968, together with Newton Heath and Patricroft, these being the last steam sheds in the Manchester area. On 30th July that year, the following withdrawn locos were in store at Bolton: 44664, 44802, 44929/47, 45046, 45104, 45290, 45312/81, 48026, 48168, 48319/80/92, 48504, 48652/92, 48702/20, together with 0-6-0 diesel-mechanical shunters D2224/6/7/34 and D2373, which had been withdrawn in April-May from Newton Heath. Two 0-6-0 diesel-electric shunters, 12012/23, were cut up in the Bolton shed yard by contractors, these having been withdrawn from 8C (Speke) and 8F (Springs Branch) respectively in December 1967.

After Bolton shed closed, 45110 moved to Lostock Hall with 45260, 48773 went to Rose Grove and 73069, which had moved to Patricroft for a few weeks during the spring, then spent its final days at Carnforth. 45110 and 48773 are still with us but 73069 did not survive into preservation, nor did 45260.

It is interesting to record that the Fowler Class 3 2-6-2T 40063 (26A), which featured on the cover of 'The Mancunian' for March 2013, spent some time in store at Bolton, although not officially allocated there. It did, however, see some use on pilot duties at Moses Gate.

Bolton shed was demolished shortly after closure. Crescent Road is still very much in existence but Back Crescent Road has disappeared and a new road named 'The Sheddings' now reminds us of former days at Bolton shed.

Sources of information:

RCTS Journal 'The Railway Observer'
Midland Railway Society – British Railways Steam Loco Shed Allocations 1950-1968
RCTS – A detailed History of BR Standard Steam Locomotives – Volumes 2 and 4
Railways in and around Bolton – Bill Simpson (Foxline Publications)
LMS Engine Sheds – Volume 3 (LYR) – Chris Hawkins and George Reeve
Bolton Engineman – Jim Markland (Foxline Publications) – 2 Volumes

25th September 1965 • Stanier Class 4P 2-6-2T No. 42626 stands at Bolton (Trinity Street) after arrival with the last train from Horwich. Photo: © Manchester Locomotive Society

September 1961 • No. 42289 simmers on Bolton shed.
Photo: D. A. Young Collection
© Manchester Locomotive Society

4th May 1968 • Stanier Class 5MT 4-6-0 No. 45110 seen under the coaling plant at Bolton shed.
Photo: D. A. Young Collection
© Manchester Locomotive Society

21st September 1961 • Stanier Class 8F 2-8-0 No. 48107 of Kettering is a visitor to Bolton shed following overhaul at Horwich Works.
Photo: I. G. Holt/ D. A. Young Collection
© Manchester Locomotive Society

20TH MAY 1962 •
Bolton resident WD 2-8-0 No. 90297 shuffles around the shed yard prior to its next duty.
PHOTO: D. HAMPSON
© MANCHESTER LOCOMOTIVE SOCIETY

A line up of locos, all withdrawn, is headed by L&Y 3F 0-6-0 No. 52443 (26D), 7F 0-8-0 No. 49592 (26A), 3F 'Jinty' 0-6-0T No. 47340 (26A), 7F 0-8-0 No. 49662 (26C) and 4F 0-6-0 No. 44000 (24G). All were dragged to Springs Branch on 3rd August 1961 for eventual cutting up at Central Wagon Co., Wigan.
PHOTO: R. GREENWOOD
© MANCHESTER LOCOMOTIVE SOCIETY

SHED CODES:
24A ACCRINGTON
24G SKIPTON
26A NEWTON HEATH
26C BOLTON
26D BURY
55C FARNLEY

15TH JUNE 1958 • Plenty of locos on view in this photo, from left to right, No. 42635 (26C), No. 45226 (24A) ex-works, No. 49674 (26C), 42630 (26C), No. 42545 (26C), No. 90699 (55C), and 49618 (26C).
PHOTO: A. GILBERT
© MANCHESTER LOCOMOTIVE SOCIETY

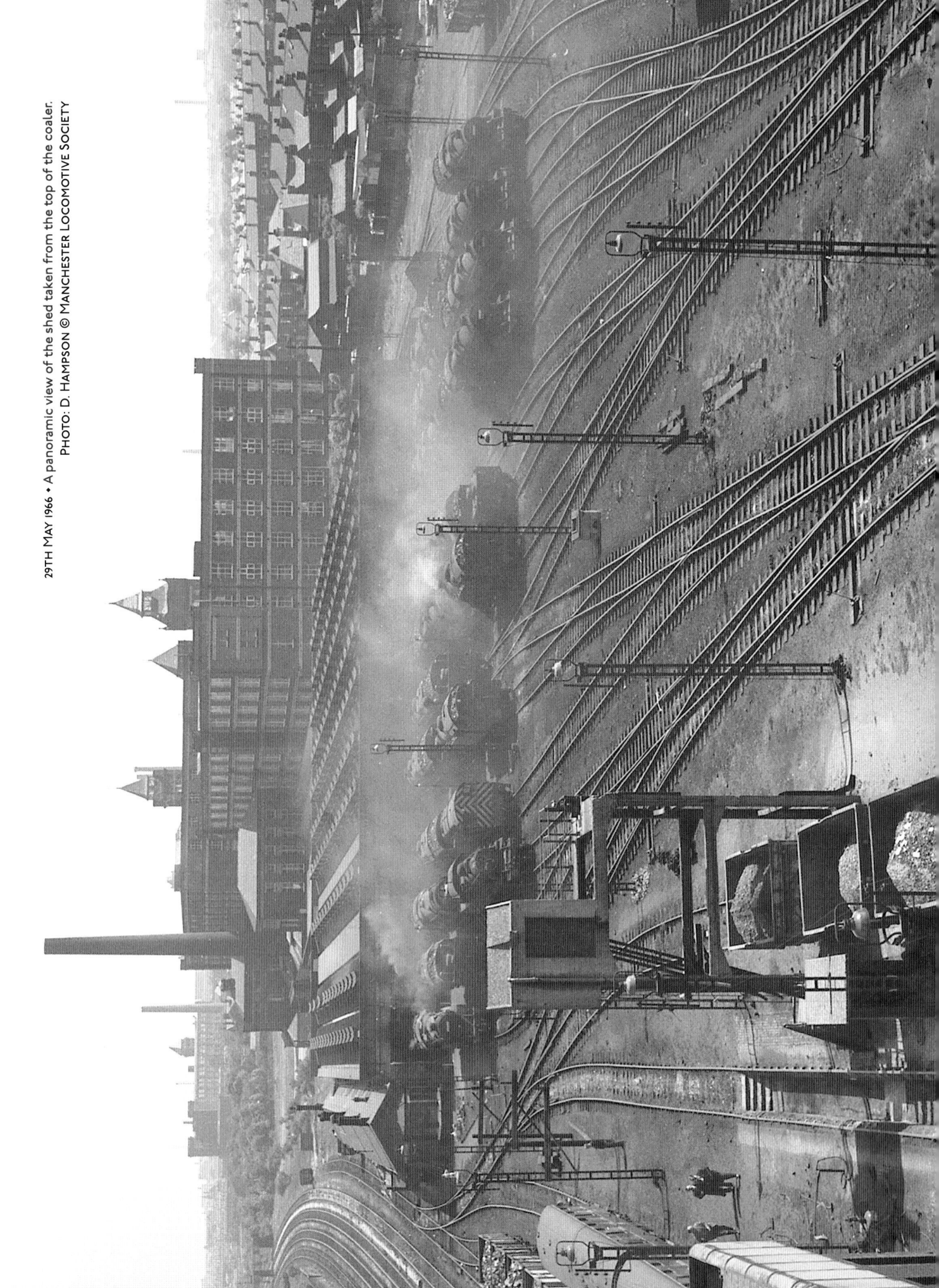

29TH MAY 1966 • A panoramic view of the shed taken from the top of the coaler.
PHOTO: D. HAMPSON © MANCHESTER LOCOMOTIVE SOCIETY

BOLTON
CIRCA 1930

18TH MAY 1968 • Although closure of the shed is not far away, there are still a healthy amount of locos on view.
PHOTO: © TRANSPORT LIBRARY

LMS-BUILT LOCOMOTIVE NAMING STRATEGY
PART 1 – BLACK FIVES AND PATRIOTS
by Philip Hellawell

2ND APRIL 1960
Fowler Patriot Class 6P5F 4-6-0 No. 45519 LADY GODIVA at Rotherham Masborough.
PHOTO: ROBERT ANDERSON/THE TRANSPORT TREASURY

Trainspotting was one of the many things my mother did not like me doing when I was a schoolboy. She felt that I should spend more time on my studies since things did not go well for me at school. Unfortunately, my inability to concentrate meant that I was never going to do well and I look back and think that, with imagination, she might have realised that from an interest in railways one could learn about geography, social history, engineering, architecture, design, and travel. The countries and states on the British Commonwealth, the monarchs of the UK, the regiments of the British Army, poets and composers, battles of the First World War, antelopes, the counties of Great Britain, literary works, and famous football clubs are just some of the name categories of steam engines.

Having said that, I have often wondered why some of the locomotive classes hinted at above were not more comprehensively named than they were. Comparatively little is known about the reasons why particular series of names were chosen or what others might have been considered.

This topic is of interest to enthusiasts who find the names of locomotives intriguing and far easier to recall than mere numbers. But historic information is not easy to find and surviving company minutes tend to show only the outcome rather than the factors that were considered.

Named LNWR locomotives were the mainstay of the LMS routes from Euston until the mid-1930s which had informative nameplates showing not only the locomotive, but the builder, location, and date. After the grouping, the LMS continued to choose names for Claughtons and Prince of Wales class locomotives intending to use the LNWR pattern of nameplate, but this never took place – presumably because of the Midland Railway's policy of not naming engines.

However, as far as this article is concerned, I have decided to restrict the scope to those locomotives built and named by the LMS itself, with slight overspill into BR days. Starting with Black Fives and Patriots, I am intending to work my way through the Jubilee, Royal Scot, Princess Royal and Princess Coronation classes.

STANIER BLACK FIVES

One of the most universally popular classes of steam engines built anywhere in the world, the LMS and later BR turned out no fewer than 842 examples of these splendid looking machines from 1934 to 1950. Suitable for almost any duty, these extremely efficient mixed traffic engines could be seen throughout the LMS system from Thurso in the north to Bournemouth in the south. The first numbered (5000) was produced by Crewe in February 1935, but Armstrong Whitworth were quicker off the mark with their first example, turning out 5020 in August 1934.

In fact, Armstrong Whitworth built the largest number, 327, with four other works being engaged in the construction, namely Crewe (241), Horwich (120, including the final 30), Vulcan Foundry (100) and Derby (54). There were innumerable detailed differences throughout the class but why did four of the Class 5s receive names? It was always a puzzle to me, and I am sure to most trainspotters of the day. Not only that but named engines were always green, weren't they? Not these – as black as the other 838.

They were not named when newly built by Armstrong Whitworth in 1935, the names being added during 1936/37.

28TH SEPTEMBER 1958 • NO. 45154 LANARKSHIRE YEOMANRY at Wakefield. PHOTO: J. E. BELL/THE TRANSPORT LIBRARY

All were based on the Northern Division at the time of naming, 5154 and 5156 at Carlisle Kingmoor, with 5157 and 5158 at St. Rollox. In 1943 the two Carlisle engines joined the other two at St. Rollox. The names and dates of naming were as follows:

45154 LANARKSHIRE YEOMANRY – 5TH APRIL 1937

The Lanarkshire Yeomanry was a yeomanry regiment of the British Army, first raised in 1819 in the Lanark area and served at Gallipoli, in Palestine and on the Western Front, following which it provided two field artillery regiments in the Second World War.

45156 AYRSHIRE YEOMANRY – 19TH SEPTEMBER 1936

The Ayrshire Yeomanry is the senior yeomanry regiment in Scotland and the seventh yeomanry regiment in Great Britain and known as The Carrick Troop. It was formed as the Ayrshire Yeomanry Cavalry in 1793, at the onset of the French Revolutionary Wars, by Archibald, Lord Kennedy. In 1838 there was a general disbandment of yeomanry regiments throughout the UK, with only two – Lanarkshire Yeomanry and Ayrshire Yeomanry being retained in Scotland, the latter serving in the South African War, at Gallipoli (Turkey), and in Palestine in 1914-18.

45157 THE GLASGOW HIGHLANDER – 6TH MARCH 1936

The Glasgow Highlander was a former Infantry Regiment of the British Army and was one of the first Territorial Regiments to go into action in the 1914-18 war, being part of the 9th Battalion of the Highland Light Infantry which landed in France in November 1914. It also saw active service in World War 2.

45158 GLASGOW YEOMANRY – 22ND MAY 1936

The Glasgow Yeomanry was first formed in 1796 as 'The Glasgow Light Horse' and disbanded in 1822, subsequently being reformed in 1848 as 'The Glasgow and Lower Ward of Lanarkshire Yeomanry Cavalry'. It was converted to an artillery unit in 1920 to form a two-battery army brigade.

Even with only four nameplates, design was not identical as *The Glasgow Highlander* had the regimental crest above the name whereas the other three had the crest below the name. Not only that, but you would have to be an eagle-eyed spotter to notice that *Ayrshire Yeomanry* had additional script in very small letters on a separate straight plate below the crest which read '*Earl of Carrick's Own*'. *Glasgow Yeomanry*, on the other hand, had the words '*Field Brigade R.A.T.A.*' on a curved plate under the crest.

3RD AUGUST 1957 • No. 45156 AYRSHIRE YEOMANRY waiting to depart from Rochdale with the 1.18pm to Liverpool Exchange.
PHOTO: © TRANSPORT TREASURY

ABOVE: SEPTEMBER 1956 • No. 45157 THE GLASGOW HIGHLANDER at its home shed, St. Rollox (65B). PHOTO: © TRANSPORT TREASURY.

RIGHT: 30TH JUNE 1963 • No. 45158's nameplate.
PHOTO: KEN NUTTALL © TRANSPORT TREASURY.

BELOW: 3RD APRIL 1961 • No. 45158 GLASGOW YEOMANRY stands at Callander at the head of an Oban-Glasgow service.
PHOTO: SID RICKARD © J & J COLLECTION/TRANSPORT LIBRARY

10TH AUGUST 1935
No. 5155 at Nottingham Midland.
PHOTO: GEORGE BARLOW © TRANSPORT TREASURY

Since Black Fives, unlike Jubilees, did not have wheel splashers, the names were mounted on back plates which were positioned above the leading driving wheels. Although none of the 191 Jubilees were built at St. Rollox, of the twelve that were first allocated to Scotland, only three were supplied with nameplates, so St. Rollox cast the other nine nameplates and those for the four Black Fives were cast and fitted using the same distinctive lettering style.

45155 QUEENS EDINBURGH – NOT APPLIED

Grammatically the name should include an apostrophe between 'n' and 's' since it is clearly meant to be possessive rather than plural, but despite this many references show the name without. It refers to the then *4th/5th Battalion (Queen's Edinburgh), Royal Scots*. This became a Royal Artillery Searchlight Regiment on 1st November 1938 whilst remaining affiliated with the Royal Scots. The 4th/5th were the first battalion of the Regiment to see action, being engaged in the first air raid of the war on 16th October 1939 when the Luftwaffe tried to bomb the Forth Bridge. On 23rd October, the 4th/5th captured the first German prisoner of war from the first enemy aircraft to be shot down. Later converted to an anti-aircraft role, on 1st November 1938 it became a searchlight Regiment of the Royal Artillery but proudly retained its Queen's Edinburgh connection.

Since it is clear that 45155 was nominated to carry this name, what went wrong? How could it be that four engines were named but with a numerical gap between the first and the second? It always seemed odd to me that 45155 remained unnamed. A batch of seven consecutively numbered engines (45153 to 45159) were still based at St. Rollox from 1948 through to c.1960 when *Ayrshire Yeomanry* and *Lanarkshire Yeomanry* were transferred to Newton Heath – a little surprising, given their Scottish names, whilst unnamed sisters 45153 and 45159 stayed on at St. Rollox.

The earliest reference to 45155 was in memoranda dated 19th February 1937 dealing with *'Naming of Class 5 Engines'*. H. G. Ivatt was involved in this correspondence, which states: "I have been informed that Major Murdoch, Second-in-Command of the Lanarkshire Regiment has been advised by Mr. Ballantyne that Engine No. 5155 could be allocated to bear the name *Lanarkshire Yeomanry*. This engine has, however, already been allocated to it the name *Queens Edinburgh* and I have suggested to Mr. Ballantyne that either Engine No. 5154 or Engine No. 5159 be chosen. On hearing further from him I will write to you again."

A further memo, seemingly, from J. Ballantyne, under the same heading reads "With reference to the above, as Engine No. 5155 has already been allocated the name of *Queens Edinburgh* I suggest that Engine No. 5154 or Engine No. 5159 be chosen to bear the name *Lanarkshire Yeomanry*. Later, the name appeared in a booklet *Modern Locomotives of the LMS* written by D. S. M. Barrie, published with authority of the LMS, which gave the names of all five named engines, including No. 5155 as *Queens Edinburgh*. Future references to this engine and its name only confuse the issue further. On 11th January 1943, The Railway Observer recorded that 5155 (12A) was seen working in Nottingham without nameplates. But in the same issue a correspondent stated that "5155 (12A) has been noted at St. Albans named *Queens Edinburgh*".

The first Ian Allan ABC of LMS locomotives of June 1943, showed the name *Queens Edinburgh* whilst a note in the 1943 Stephenson Locomotive Society Journal said 5155 had been seen without a name. Another reference in the November/December 1944 issue of The Railway Magazine said: "LMS 4-6-0 No. 5155 *The Queens Edinburgh* and some other locomotives have had their names removed." Despite all these conflicting reports, no photographic evidence has ever been produced that 5155 carried a nameplate, so the matter will always be open to conjecture.

Patriots

Predecessors, the Claughton Class, had been 130 strong when built by the LNWR between 1913 and 1921, of which 60 were named, and they all passed into LMS ownership in 1923 – the former *Princess Louise* lasting long enough (1949) to achieve its BR number of 46004. In 1929, the first Claughton to be withdrawn was No. 5971 which, hauling a sleeping car express from Euston to Glasgow, collided head-on with a 4F at Doe Hill on a Sheffield-Birmingham freight which was wrong line. Leeds-based Claughton, 5902 was also involved in a head-on accident in 1930 with a 4F which had been shunted into its path. Both these engines were beyond economic repair, but salvageable parts were to form part of the construction of the first two Patriots.

Between the introduction of the 'Royal Scot' class 4-6-0s in 1927 and the first of the Stanier Pacifics in 1933, the LMS built the 'Baby Scots'. Always my favourite class of engine type due to their square lines the class came to be called Patriots. Introduced towards the end of the Henry Fowler regime, these elegant 4-6-0s with parallel boilers were mainly built at Crewe (40), but Derby also contributed 12, including the first two engines which were classified as 1930 rebuilds of the LNWR 'Claughton' 4-6-0s and used their coupled wheelsets (after modifications), bogies, axleboxes, brake gear, reversing screws and whistles. They could always be distinguished from the rest of the class by the larger centres of the driving wheels and the fluted coupling rods.

The next 38 were considered to be renewals of Claughton engines, and were allocated numbers from the Claughton range even though use of any of their parts was minimal. The final ten locos were classified as new engines which gave a total of 52 in main line service between 1932 and 1934 which could be seen at work over almost the whole of the LMS system on a wide variety of duties – they went everywhere and did everything.

Far from having a policy, this class was an unfortunate target for random naming and numbering, initially developing through replacements – for each Claughton withdrawn, a Patriot bearing the same number entered service. The first withdrawal to be treated in this way, 5971, had been named *Croxteth* but its replacement emerged from Derby in late November 1930 unnamed. The next one however, 5902 *Sir Frank Ree* entered service a few days later, with a nameplate not unlike those on the Royal Scots.

From then on until July 1931 these were known as Rebuilt Claughtons. But in the space of the next fourteen months to September 1933 a further forty appeared. They, neither looked nor sounded like Claughtons, being far more like Royal Scots. By chance only two of the next twenty two to be converted had previously carried a name and, of these the ninth *Bunsen* and, the fourteenth *Lady Godiva*, like the prototype *Croxteth* went into service nameless so, by February 1933, only 5902 *Sir Frank Ree* of the growing class, now totalling 24 locomotives, was named.

However, in March 1933 the name *Sir Frederick Harrison* entered service on loco 6027 from Crewe with *E. Tootal Broadhurst* similarly adorning 5935 from Derby. All the rest entitled to a name got one. Moreover, the names of *Bunsen*, *Lady Godiva* and *Croxteth* were restored to their rightful owners in the latter part of 1933. At this time, the class totalled 42, of which twelve were named. In 1934 those original 42 were renumbered in chronological order of building, 5500 to 5541 to precede the numbers of the additional ten entirely new engines built at Crewe in 1934 which had been allocated the numbers 5542 to 5551. Five more of the class, to be numbered 5552 to 5556 were on order, but they were instead built by Stanier in 1934 with taper boilers and top feeds, becoming the first of the Jubilee class.

However, as the Claughtons had been disappearing from

16th June 1952 • No. 45500 Patriot at Huddersfield working the 10.50 to Liverpool. The large centre bosses of the driving wheels and the fluted coupling rods from its Claughton predecessor are clearly visible. Photo: EB © Transport Treasury

service so, for the most part, did their names. Two names which survived were those of the two LNWR employees who had been awarded the Victoria Cross, but that of *Patriot* did not. After something of a campaign coupled with the LMS not liking the soubriquet 'Baby Scots,' it was decided that the first loco conversion, by this time named *Croxteth*, should be renamed *Patriot*, which it was on 25th February 1937, the opportunity being taken to designate the whole series 'Patriot Class'.

Presumably, it must have been the intention to name every engine in the class. Those that were received a perplexity of differing names apart, mystifyingly, from 10 engines which remained nameless even though names had been allotted but never applied. Fourteen names were initially perpetuated from the Claughton class, and some were historical names removed from Royal Scots. Others were naval and military personnel, regiments of the British Army, railway directors, a school for the blind, a public school, plus, imaginatively, the names of popular holiday resorts or destinations served by the LMS along the Lancashire and North Wales coasts. Of this latter category, six also carried the appropriate civic coats-of-arms above the nameplates.

Also, there had been some transferring of names, within the class to form groupings of military, resorts, and directors although even then it was haphazard. The 34 engines keeping their original form were substantially unaltered during their 30 years or so of service. They were all coupled with Fowler 3,500 gallon, 5½ ton coal capacity tenders, although the high-sided 7 ton coal capacity version did make an odd appearance on a couple of locos.

Two original locos lasted until November 1962 – 45543 and 45550, as such they were among the last main line passenger locomotives to have their origins in pre-grouping days. All the taper-boilered engines bore names at the time they were rebuilt with just one exception, but I am going to deal with the original ones first and follow on with the rebuilt ones. In the headings, I use the last (i.e. BR) number series to identify each locomotive.

45500 Patriot

This loco was first designated as a rebuilt Claughton and numbered 5971 *Croxteth*, becoming 5500 in 1934 and acquiring its

19TH MAY 1958 • No. 45501 St. Dunstan's at Camden.
Photo: Transport Treasury

BR number in April 1949. In 1937, it took on the mantle of the name *Patriot*, but not the actual nameplate, from Claughton No. 5964. This had been the LNWR war memorial engine, outshopped in black, originally numbered 69, then 5964, and, finally, 1914 in recognition of its role.

During the 1950s, 45500 was suitably and respectfully bedecked with poppies and garlands for the 11th November annual Remembrance Day event at Rugby engine shed, a role later taken over by Royal Scot 46170 *British Legion*. I always think what a shame it was that a locomotive which commemorated the war dead of the LNWR was not considered for preservation, especially as an example of a class which was a major stepping stone between traditional LNWR designs and hugely successful LMS successors.

45501 St Dunstan's

Like 45500 this was, for accountancy purposes, considered to be a rebuilt Claughton, this time from 5902, the number allocated when built in 1930. Renumbered as 5501 in April 1934, it was originally named *Sir Frank Ree*, but this was changed to *St. Dunstan's* in 1937 and finally numbered 45501 in March 1949. It was reallocated 23 times during its 31 years of service covering over 1¼ million miles. It spent time at Crewe, Camden, and Longsight sheds, amongst others, before finishing its days at Carlisle Upperby in 1961.

St. Dunstan's was named after the London charity, now known as Blind Veterans UK, established in 1915 to help those many veterans who lost their sight during the First World War. The naming ceremony was performed at Euston on 17th April 1937 by the war-blinded MP, Sir Ian Fraser CBE. The nameplate was striking and unique, being a brass replica of the St. Dunstan's badge comprising a Grecian torch surrounded by an oval shield with a simple banner at an angle across the centre bearing its name.

45502 ROYAL NAVAL DIVISION

Later 63rd (Royal Naval) Division, it was formed at the outbreak of the Great War when it was realised that the Royal Navy had an excess of recruits, and not enough jobs for them. They were initially made naval reservists but reorganised into the nucleus of an Army-style division of light infantry for deployment on land. This was named the Royal Naval Division which served with distinction both at Gallipoli and on the Western Front.

The loco was named at Euston by Winston Churchill on 5th June 1937, this engine was the first of the class to be withdrawn in September 1960. Despite that fact, it achieved the second-highest mileage of any of the original examples at 1,388,595, over 200,000 more than 45549 which had the lowest total despite having been in service for 21 months longer.

APRIL 1959 • No. 45502 ROYAL NAVAL DIVISION works a fitted freight at Farington.
PHOTO: BILL ASHCROFT © THE LMS-PATRIOT PROJECT

45503 THE (ROYAL) LEICESTERSHIRE REGIMENT

This infantry unit was raised in 1688 and subsequently served in many British Army campaigns during its long history. The 1st Battalion spent the whole of the First World War on the Western Front, arriving in August 1914 and going on to fight in many of the major battles of the campaign. In October 1914, it was joined by 2nd Battalion, which remained in France and Flanders until late 1915 and then served in Egypt, Mesopotamia, and Palestine, where it assisted in the final defeat of the Turks.

In 1946, the regiment was given the 'Royal' prefix in honour of its extensive wartime service. Originally named *The Leicestershire Regiment*, the locomotive was similarly renamed on 3rd November 1948, the original regimental crests being retained. It spent much of its life at Crewe, ending its days at Carlisle Upperby and being withdrawn in August 1961

22ND MARCH 1953 • No. 45503 THE ROYAL LEICESTERSHIRE REGIMENT outside Crewe North Junction signal box.
PHOTO: JIM FLINT/JIM HARBART © THE TRANSPORT TREASURY

28TH JANUARY 1947 • A works photo of No. 45504 ROYAL SIGNALS with sans serif cabside numbers and tender lettering, the smokebox retains the scroll-type numerals. PHOTO: MILEPOST 92½ © THE TRANSPORT TREASURY

45504 ROYAL SIGNALS

A combat support arm of the British Army, it is the Army's expert engineering and communications operator, dealing with information systems, networks, power plants, and cyber techniques. Royal Signals units provide the digital backbone for the British Army, wherever it operates in the world. They have deployed on every operation the Army has been involved in, home and overseas, with special forces and intelligence gathering teams, or serving in armoured formations.

Passing through Willesden, Crewe North, and Carlisle, it was one of three locos attached to Bristol Barrow Road (i.e. Western Region) in 1958 to work cross-country expresses to Birmingham, Leeds, and York. The trio were also used on fitted freights to the North and even into South Wales on coal trains.

45505 THE ROYAL ARMY ORDNANCE CORPS

The Royal Army Ordnance Corps was responsible for procurement, storage and the issue of weapons, ammunition, and fortifications, supplying everything the British Army needed in peace and war. My father served in this army division throughout the Second World War, being based near the Suez Canal in Palestine for five years. The long history of RAOC stretches back to at least 1299 when the Keeper of the Kings Wardrobe in the Tower of London was given responsibility for Warlike Equipment and Military Expenditure. In 1993 it became part of the new Royal Logistic Corps.

Named in 1947, this loco was based at Rugby, Camden, Crewe, and Carlisle at times, ending its days at Lancaster in June 1962. For a period, it was one of two paired with a straight high-sided 3,500 gallon tender with 7 tons coal capacity.

45506 THE ROYAL PIONEER CORPS

Formed on 17th October 1939, as The Auxiliary Military Pioneers Corps, it was re-designated Pioneer Corps in 1940 and gained the Royal suffix in 1946 recognising its valuable work during the war. Many of those who served were either veterans, too old to be on the front line, or had medical issues. Its recruits included Jewish and anti-Nazi refugees who had fled from Austria, Germany, and Eastern Europe. The Corps was ideally suited where manual labour was required, such as railway maintenance, light engineering tasks and mine clearance.

This was one of the three parallel boiler Patriots transferred to Bristol Barrow Road in 1958. It remained there until withdrawal in 1962, often finding its way to Bath, Birmingham, and York

45507 ROYAL TANK CORPS

Tanks were first used in September 1916 during the Battle of the Somme in World War 1. Now called the Royal Tank Regiment, it is the oldest tank unit in the world. The Battle of Cambrai (1917) showed the true potential of combined-arms manoeuvre, when tanks, infantry, artillery, engineers, and aircraft all worked together to shatter German defences. In its short life the Corps had grown to 26 battalions and was operating Mk IV and Mk V heavy tanks, Whippet light tanks, and armoured cars.

Built at Crewe in August 1932 and named in November 1937, it spent much of its life at Rugby, Carlisle Upperby and Preston. Towards the end of steam, it was one of a batch of Patriots ending their days on Morecambe–Lancaster–Leeds trains, before withdrawal from Lancaster in October 1962. Fittingly, one of the nameplates with crest is displayed in the Tank Museum at Bovington in Dorset.

JULY 1960 • No. 45505 THE ROYAL ARMY ORDNANCE CORPS paired with a high-sided tender, pulls into Bangor station.
PHOTO: DON MATTHEWS
© THE TRANSPORT TREASURY

SEPTEMBER 1953 •
No. 45506 THE ROYAL PIONEER CORPS passes through Colwyn Bay station.
PHOTO: ROY EDGAR VINCENT
© THE TRANSPORT TREASURY

13TH JULY 1957 • No. 45507 ROYAL TANK CORPS on the 11.33am Lancaster-Glasgow Central leaving Carstairs. PHOTO: W. A. C. SMITH
© THE TRANSPORT TREASURY

27TH FEBRUARY 1960 • No. 45509 THE DERBYSHIRE YEOMANRY and Black Five No. 45156 AYRSHIRE YEOMANRY pass through Church Fenton with empty parcels stock from York to Manchester.
PHOTO: MIKE MITCHELL © THE TRANSPORT TREASURY

45509 THE DERBYSHIRE YEOMANRY

First raised in 1794, it served as a cavalry regiment and dismounted infantry regiment in World War 1 and provided two reconnaissance regiments in the Second World War. In 1957, it amalgamated with the Leicestershire Regiment to form the Leicestershire & Derbyshire Yeomanry.

For almost 20 years, this loco was unnamed but, in November 1951 it became the only original Patriot to be allocated to Derby after nationalisation, where it was named *The Derbyshire Yeomanry* upon transfer from Crewe. It remained there for the next seven years until August 1958 when it was transferred to Newton Heath depot and often seen, usually double heading, on the afternoon Heaton to Red Bank Parcels down the Calder Valley, returning the day's empty newspaper vans to Manchester. Regarded as a poor steamer, it was the only member of the class to have been damaged by enemy action during the war. One of its nameplates is displayed at Derby City Museum.

2ND JUNE 1957 • No. 45511 ISLE OF MAN at Euston with the R.C.T.S. 'The Mercian' railtour and a close-up view of the nameplate.
BOTH PHOTOS: A. E. BENNETT © THE TRANSPORT TREASURY

45511 ISLE OF MAN

This was the first resort naming and carried the island's crest above the nameplate. Built at Crewe Works in August 1932 and first bearing LMS Claughton number 5942, it was renumbered 5511 in August 1934 and 45511 in May 1949. It saw no less than 22 transfers, before being withdrawn from Carlisle Upperby in February 1961

The island, annexed by England in the 13th century and ruled by a Lieutenant Governor appointed by the Crown but with its own Parliament, was my late wife's and my favourite holiday destination, despite us having travelled all through Europe and many other world countries. Having holidayed there on a few occasions since the age of 10, I am lucky to have travelled over every inch of the steam railway system apart from the 2½ mile Foxdale branch.

Before the package holiday era, the Isle of Man was a popular holiday destination with the slogan *"Go abroad to the Isle of Man"*. Sea crossings were operated by the LNW, L&Y and Midland railways plus the Isle of Man Steam Packet Company from Liverpool, Fleetwood, and Heysham, principally to Douglas, very occasionally to Ramsey.

45515 CAERNARVON

This name was first allocated to LNWR 'George V' Class 4-4-0 No. 984. The Patriot was delivered from Crewe Works in October 1932 with LMS No. 5992, renumbered 5515 in August 1934, named in 1938, and gained its final number in April 1948. The loco spent most of its life at Aston depot until withdrawal from Newton Heath in June 1962.

Located overlooking the southern end of the Menai Straits on the former LNW and LMS line between Bangor and Pwllheli, Caernarvon, the ancient capital of Wales, is famous for its impressive castle, the foundation stone of which was laid in 1283. In 1969, it was the setting for the investiture of Prince Charles as Prince of Wales.

FEBRUARY 1950 • No. 45516 with members of THE BEDFORDSHIRE AND HERTFORDSHIRE REGIMENT at Southampton.
PHOTOS: DAVID ANDERSON © THE TRANSPORT TREASURY

45516 THE BEDFORDSHIRE AND HERTFORDSHIRE REGIMENT

This, with 39 letters, was the longest name of the Patriots, bestowed on 31st July 1938. Its most momentous day occurred in February 1950 when it hauled a special troop train for its namesake regiment from Southampton Docks along the L&SWR main line.

The Bedfordshire and Hertfordshire Regiment was the final title of a Line Infantry regiment of the British Army, originally formed in 1688. After centuries of service in many small conflicts and wars, including both the First and Second World Wars, the regiment was amalgamated with the Essex Regiment in 1958 to form the 3rd East Anglian Regiment (16th/44th Foot). However, this was short-lived, a further amalgamation in 1961 forming the present Royal Anglian Regiment.

45518 BRADSHAW

This engine was named *Bradshaw*, briefly from 1939, then permanently in October 1947 after the English cartographer, printer and publisher George Bradshaw (1801 to 1853). Bradshaw's Guide produced by Bradshaw and his successors was the longest-running and most successful series of publications in the world covering railway timetables and tourist information in Great Britain and abroad.

Built at Crewe in 1933, it served at Camden, Willesden, Carlisle Upperby and Edge Hill before withdrawal in October 1962 when it was cut up at Horwich works.

45519 LADY GODIVA

Named after 11th Century Anglo-Saxon Countess of Mercia Godiva, whose husband, Leofric, Earl of Mercia, growing exasperated over her pleas to reduce Coventry's heavy taxes, declared he would do so if she rode naked on horseback through the crowded marketplace. Legend has it she did exactly that with

26TH OCTOBER 1961 • No. 45515 CAERNARVON (26A) at Lancaster Castle with 2.30pm Morecambe to Crewe (1K78).
PHOTOS: © RON HERBERT/THE LMS-PATRIOT PROJECT

20TH JUNE 1959 • No. 45518 BRADSHAW passes through Cheddington station. PHOTO: DONALD ROBERTSON © THE TRANSPORT TREASURY

her long hair able to cover her modesty. An impressive statue by William Reid-Dick was presented to the city in 1949 to boost morale and symbolise the regeneration of Coventry after its bombing.

One of the three allocated to Bristol Barrow Road, along with 45504 and 45506. All, oddly, were among the six Patriots attached to the same tender throughout. The nameplates of *Lady Godiva* were presented to the Herbert Museum in Coventry in 1962 *(see photo on page 24)*.

45520 LLANDUDNO

The name of this engine carried the civic coat-of-arms crest, and it is believed that the ceremony took place at Llandudno station in 1937. Built at Derby Works in 1933, and numbered 5954, it was renumbered 5520 in September 1934 and given its final number in July 1948. Unusually, *Llandudno* spent most of its time at two depots, Longsight and Edge Hill, from where it was withdrawn in May 1962.

21ST FEBRUARY 1953 • No. 45520 LLANDUDNO shortly after departure from Manchester London Road.
PHOTO: B.K.B. GREEN. © MANCHESTER LOCOMOTIVE SOCIETY

Arguably the most popular of all the Welsh coast holiday destinations, the resort is a convenient location for holidaymakers from the North and Midlands. Due to the easy journey, the LNWR provided it with club trains from 1908 to carry businessmen to their office jobs in Manchester. For those paying a supplement to join the club, special saloon carriages offered comfort and privacy. They continued to operate under the LMS between the wars but were discontinued during World War Two and not resurrected afterwards.

45524 BLACKPOOL

The loco was built at Crewe Works in March 1933 and named *Sir Frederick Harrison*, which name was transferred to 45531 in 1937. It was finally renamed *Blackpool* at the town's station on 22nd March 1937 in the presence of Sir Josiah Stamp, the LMS Chairman. *Blackpool* regularly worked the LMS' 'Fylde Coast Express' between London Euston and Blackpool and was withdrawn from Liverpool Edge Hill in January 1964.

Developing early as a holiday destination for the workers of the North and Midlands of England plus Scotland, Blackpool was very popular in the summer holidays but, shrewdly, the town extended its season from 1879 onwards with the famous annual 'Golden Mile' of illuminations. At first reached by a branch line of the Preston and Wyre Railway, it has a seven mile promenade dominated by the 518 feet high Blackpool Tower, built in 1894.

Moreover, Blackpool became the first town in the UK to have a permanent electric tramway. Not restricted to public transport, open-top trams were a captivating part of the holiday experience packed with tourists enjoying the sun and sights. Although, in the mid-20th century, the rise of motor transport had led other British towns and cities to abandon their tramways, Blackpool's visionary decision to retain and operate theirs has made it an integral part of the town's identity.

Railway lines into Blackpool were jointly owned by the L&Y and the LNW, their termini being Blackpool Talbot Road (later North) and Central. North station was rebuilt with 15 platforms in 1898, whilst Central station had 14 platforms after rebuilding in 1900. By 1903, the original coast line from Lytham was bypassed by a direct line from Kirkham, which gave a five mile shorter route from Preston. One Saturday in August 1935 saw the remarkable number of 467 trains arriving and departing from the resort's stations.

The advent of motor transport saw the closure of Central station in 1964 and the closure of the direct line in 1967. The coast line was truncated at Blackpool South, all main line services were then concentrated at Blackpool (North), the station being rebuilt in 1974 into the modern terminus we see today.

An undated shot of No. 45524 BLACKPOOL departing Crewe station.
PHOTO: J. SPENCER © MANCHESTER LOCOMOTIVE SOCIETY

45533 LORD RATHMORE

The name was intriguing, being one of only two Patriots with links to the House of Lords. Rathmore is a town in County Kerry, whilst David Robert Plunket (first Baron Rathmore), born 3rd December 1838, was a barrister at King's Inn, Dublin, later a QC, and Conservative MP for Dublin University, a seat he held for 25 years. The peerage became extinct in 1919 on his death. However, his name endured for another 42 years on 45533, until the autumn of 1962.

Bearing in mind his date of death, 22nd August 1919, why was 5533 was put into traffic in April 1933 bearing his name? The LNWR began building Claughtons in 1911 with an initial batch of 20 entering service in 1913 and 1914. Nineteen were named after directors and senior officers of the company, one of these being Lord Rathmore, who was also a director of the Suez Canal Company. The name was transferred to this Patriot when built.

From 1948 to 1958 *Lord Rathmore* was a Liverpool Edge Hill (8A) engine, with moves to Carlisle, Rugby and Nuneaton following. The loco finished its days at Edge Hill in September 1962 and is fondly remembered by the author passing through Sowerby Bridge on 15th April 1961 on an early evening return working carrying St. Helens Rugby League supporters back to Lancashire. Their team – 26-9 victors over Hull F.C. – had been playing in a R.L. Cup semi-final at Odsal Stadium, Bradford, a walkable distance from Low Moor station.

45537 PRIVATE E. SYKES V.C.

Another of the Claughton class names that was transferred to a Patriot, the engine entered service in July 1933 with its transferred name already in place. A fifteen year Preston resident, it was withdrawn in 1962.

Ernest Sykes was born in Mossley in 1885. When war broke out, he was a trackman for the LNWR. Age 29 when he joined up with the Duke of Wellington's Regiment (West Riding). He was

No. 45533 LORD RATHMORE at Preston with the 4.20pm local working to Wigan.
PHOTO: © BILL ASHCROFT/ THE LMS-PATRIOT PROJECT

5TH APRIL 1961 • No. 45537 PRIVATE E. SYKES V.C. at Syston East Junction on the 3.45pm Peterborough to Leicester local.
PHOTO: © FRANK CASSELL

4TH NOVEMBER 1938 • Crowds gather at Settle station to witness the naming ceremony of No. 5538 GIGGLESWICK.
PHOTO: COURTESY: THE LMS-PATRIOT PROJECT

badly wounded at Gallipoli in 1915, nearly losing a foot. But, on 19th April 1917, Private Sykes won his Victoria Cross at Arras while serving with the Northumberland Fusiliers. The citation reads that he: *"went forward and brought back four wounded men. He made a fifth journey and remained out under conditions which appeared to be certain death until he had bandaged all those who were too badly wounded to be moved. These gallant actions performed under incessant machine gun and rifle fire showed utter contempt of danger"*.

Ernest Sykes returned to his work on the railways after the war and the LNWR fittingly named a 'Claughton' class engine after him. Serving in the Home Guard during the Second World War, he died in 1949, just before he was due to retire. His Victoria Cross is on display at the Northumberland Fusiliers Museum at Alnwick Castle.

45538 GIGGLESWICK

Giggleswick is an ancient and attractive village in rural North Yorkshire with a population of 1,270 about a mile from Settle. However, the locomotive was named not after the village but, intriguingly, after the public school which bears its name. Russell Harty, a hugely popular TV chat show host from 1972 to 1988 was a teacher of English at the school from 1958 to 1964 and, from a railway perspective, one of its most prominent former pupils was O. S. Nock, a prolific and highly respected author of numerous books and articles concerning railways.

It happened that, on 2nd July 1938, Vice-President of the LMS, Sir Harold Hartley FRS was chief guest at Giggleswick School annual speech day. After presenting the prizes he said that he thought it could be arranged to give one of the unnamed Leeds-based Patriots the name *Giggleswick*. 5538 was the chosen loco and the public naming ceremony of the immaculately turned out engine in LMS crimson livery took place at Settle station on 4th November 1938. Spending twelve years at Holbeck depot in Leeds, *Giggleswick* undertook its last operational tasks at Nuneaton in 1961, was withdrawn in 1962, and cut up at Thomas Ward's in Sheffield.

45539 E. C. TRENCH

Ernest Frederic Crosbie Trench CBE, TD (6th August 1869 to 15th September 1960) was a British Civil engineer. Working primarily on the railways, he was appointed the chief engineer of the LMS in 1923. He was a member of the Institute of Civil Engineers from 1897 and served as its president from 1927-1928, retiring on 1st April 1930. In 1920 he was awarded the CBE in recognition of his war service and in 1931 received the Territorial Decoration for his work as a Volunteer Colonel in the Engineer and Railway Staff Corps.

The loco had long spells of five years Aston, nine at Crewe and six at Longsight.

22ND AUGUST 1952 • No. 45539 E. C. TRENCH at Stockport with an Up service. Over four years after its demise LMS still adorns the tender. PHOTO: © E. R. MORTEN/MANCHESTER LOCOMOTIVE SOCIETY

No. 45541 DUKE OF SUTHERLAND in early British Railways livery on Polmadie shed, Glasgow on an unrecorded date.
PHOTO: © THE TRANSPORT TREASURY

45541 DUKE OF SUTHERLAND

Named after Cromartie Sutherland-Leveson-Gower KG, the 4th Duke of Sutherland who died in 1913, having been a Director of the LNWR for 38 years. He was at one time MP for Sutherland. His grandfather, the 2nd Duke, had been a Director of the Liverpool and Manchester Railway, and his father the Marquis of Stafford had joined the LNWR Board in 1854.

Only two Patriots were named after people with links to the House of Lords, this being one of them. It spent most of its life at Longsight and Carlisle Upperby.

45543 HOME GUARD

Formerly known as the Local Defence Volunteers it was, at first, a rag-tag militia, with scarce and often make-do weapons and uniforms. Yet it evolved into a well-equipped and well-trained army of 1.7 million men. Prepared for invasion, it also performed roles including bomb disposal and manning anti-aircraft and coastal artillery. Over the course of the war 1,206 men of the Home Guard were killed on duty or died of wounds.

The engine was outshopped from Crewe on 16th March 1934 and, on 30th July 1940, it received its name in a ceremony at Euston, officiated by Lieutenant General Sir Henry Pownall KBE, CB, MC, Inspector General of the Home Guard. This was the first naming following the outbreak of war.

45546 FLEETWOOD

This loco was one of the final series of engines of the class built at Crewe Works in March 1934 and named by the LMS in 1938. Possibly the only example to be allocated to Carlisle Kingmoor, it worked trains over the Settle-Carlisle line for a period. Withdrawn from Warrington shed in 1962, it recorded the second lowest mileage of any of the class at 1,212,897.

Famous as one of Britain's chief fishing ports, Fleetwood was a town created by the railways. Initially a series of sand dunes, it was developed in the 19th-century by landowner Peter Hesketh-Fleetwood. In 1835, he realised that, by establishing a port and connecting it by railway, he could capture trade from Liverpool.

4TH AUGUST 1962 • No. 45543 HOME GUARD pictured at Skew Bridge with a local service for Preston. PHOTO: BILL ASHCROFT © THE LMS-PATRIOT PROJECT

A large group of 'spotters' watch as No. 45546 FLEETWOOD powers its freight working through Lichfield on an unrecorded date.
PHOTO: © THE TRANSPORT TREASURY

London architect Decimus Burton designed the town for Fleetwood who stipulated that it had to bear his name. In 1840 the first passengers arrived along the Preston & Wyre Joint Railway and, with it being the first resort which Lancashire workers could get to by train, there was an incredible 100,000 visitors in the opening year.

At this time there was no railway past the Lake District, so the plan also envisaged passengers travelling to Fleetwood by rail and then embarking on a boat to Scotland. To service this market, Hesketh-Fleetwood designed and had built the North Euston Hotel, an impressive semi-circular building, still in business today, and boasting a Grade 2 listing from English Heritage.

By 1841, steamer services were operating to the Isle of Man, Whitehaven, Belfast, and Ardrossan for Glasgow. It later developed into an important base for the Lancashire & Yorkshire Railway, operating cross-channel steamers from its port to Belfast and the Isle of Man, an impressive stone plinth which stands in the park on the promenade commemorates the L&Y's

LMS Patriot No. 5905 LORD RATHMORE is seen passing Elstree with a St. Pancras to Leeds express in 1934. In September 1934 the number was altered to 5533, the name remained unchanged. During the BR period as No. 45533 it had a long stint allocated to Edge Hill, Liverpool which ended in 1959 with a transfer to Carlisle Upperby, from where withdrawal took place in September 1962.

operations. Sailings still leave from Fleetwood to Larne twice a day with an occasional sailing to Douglas. The 1841 ferry from Fleetwood to Knott End, a pleasant seven minute trip across the River Wyre still operates today, saving a 13 mile road journey.

45548 LYTHAM ST. ANNES

Built by the LMSR at Crewe Works in May 1934, this was numerically the last of the resort-named 'Patriots' and also the last, numerically, to carry any name. The engine was named by the town's mayor at the station on 18th December 1937 and was one of six members of the class to receive a civic coat-of-arms crest above the nameplate. It was one of only 6 class members to be paired with the same tender throughout its lifetime and ended its days at Nuneaton from where it was withdrawn in June 1962.

Considered more refined than neighbour Blackpool, the town is another of the popular holiday destinations at one time served by the LMS and is also well-known for its championship golf courses, notably Royal Lytham St. Annes.

No. 45548 LYTHAM ST. ANNES pictured at Skew Bridge, Preston, date unknown. PHOTO: BILL ASHCROFT © THE LMS-PATRIOT PROJECT

REBUILT PATRIOTS

As a result of the success of rebuilding two Jubilee class locomotives, 45735 *Comet* and 45736 *Phoenix*, H. G. Ivatt completed the rebuild of eighteen Patriots over a 2½ year period from 1946 as more Class 6 power was required after World War 2. The frames were kept, but they were fitted with larger Stanier Type 2A taper boilers, new cylinders, Stanier cabs (unlike the rebuilt Royal Scots), double chimneys and were coupled with 4,000 gallon, 9 ton coal capacity Stanier tenders. Not immediately, but starting in 1948, they were also fitted with Royal Scot-type curved smoke deflectors. They were virtually identical to Stanier's rebuilt Royal Scots and the two rebuilt Jubilees.

Although the end of steam was not in sight, there never was a plan to reboiler more than 18 Patriots, and the remaining 34 examples, thankfully for those of us who preferred that design, retained their parallel boilers until the end. The rebuilt locos were either outshopped from the works already named or received their names very shortly afterwards, with one exception, as we shall see.

The Patriots chosen for rebuilding were simply the next ones due for General Repair, although none of the first twelve were rebuilt due to 'non-standard features'. Those that were rebuilt were considered to be the equal of the rebuilt Royal Scot Class and were said to be much better riding engines.

BELOW: 13TH JULY 1963 • No. 45512 BUNSEN passing Carstairs on 1M58, a Glasgow Central-Blackpool special.
PHOTO: W. A. C. SMITH © THE TRANSPORT TREASURY

45512 BUNSEN

Of all the names allocated to the Patriots, this was the biggest puzzle to me in my trainspotting days. Named after the German chemist and inventor of the Bunsen Burner, Robert Wilhelm Bunsen (1811-1899). Never married, he lived for his students, with whom he was very popular, and his laboratory. Among many achievements, he discovered Caesium and Rubidium and also found an antidote to arsenic poisoning..

Turned out by Crewe in September 1932, the loco was rebuilt in January 1948. It had the highest mileage of any of the class, having travelled 1,769,977 miles in service, many of these whilst it was allocated to Carlisle Upperby between 28th May 1949 and being one of the last to be withdrawn, on 27th March 1965. This makes it the longest period that any member of the class was allocated to one depot.

45514 HOLYHEAD

This engine was built at Crewe Works in September 1932 and was named in 1938 although no official ceremony of the occasion has been recorded. Originally numbered 5983, it became 5514 in August 1934 and 45514 in May 1948. *Holyhead* was rebuilt in March 1947 and was shedded at Holyhead for a couple of short periods in May 1947 and August 1948 but spent most of its working life at Camden depot before being the first of the rebuilt types to be withdrawn by nearly two years, which it was from Derby shed in May 1961.

Situated on Holy Island off the coast of Anglesey, the port of Holyhead was the western terminus of the LNWR, the City of Dublin Steam Packet Company and the LMS for their cross-

1960/61 • No. 45514 HOLYHEAD at Wellingborough on the Midland main line. PHOTO: BARRY RICHARDSON © THE TRANSPORT TREASURY

channel Irish passenger, freight, and mail services to and from Dublin. It was also the ultimate destination for the famous train *The Irish Mail* from London Euston whose restaurant cars were famed for the quality of their catering.

45521 RHYL

The popular North Wales coast was a favoured destination in LNWR and LMS days, with many profitable excursion trains from the Lancashire and Yorkshire towns at bank holidays and during their annual 'Wakes Weeks.'

Rhyl was Derby-built in March 1933, named in 1937 and carried a civic coat-of-arms crest above the nameplates, both of which have survived in preservation. The second member of the class to be rebuilt, in November 1946, *Rhyl* spent 14 years at Edge Hill depot from 1947, working trans-Pennine expresses. It was withdrawn from Wigan Springs Branch shed in September 1963.

45522 PRESTATYN

This was the last of the 18 Patriots to be rebuilt with a Stanier Type 2A taper boiler. Originally built at Derby, in 1933, *Prestatyn* received its name in 1939, probably at Prestatyn station. It emerged in its new form from Crewe Works in January 1949 and, before its withdrawal from service at Manchester Longsight depot in September 1964, it had been based at Bushbury, Crewe North and Camden.

One of the popular North Wales coastal holiday resorts, Prestatyn, like its neighbours, Rhyl and Colwyn Bay was known for its miles of sandy beaches and visitor facilities.

45523 BANGOR

Bangor was built at Crewe Works in March 1933 as No. 6026, and renumbered 5523 in 1934, named in 1938, and rebuilt in October 1948, before finally being numbered 45523. Its withdrawal came in January 1964 after main line service at Bushbury, Crewe North, Camden and finally Willesden.

With a cathedral and university, Bangor is an important cultural North Wales city, one of the smallest in the UK. Situated on the Menai Straits overlooking the Menai Bridge and the Isle of Anglesey, it is home to the National Trust-owned Penrhyn Castle, two miles to the east, which long had a display of early steam locomotives. Now, only those which relate to Penrhyn Quarry have been retained, the others having quite recently been dispersed into the preservation movement.

45525 COLWYN BAY

Built at Derby Works in 1933, it was originally numbered 5916 and named *E. Tootal Broadhurst* and was officially renamed *Colwyn Bay* at the town's railway station on 16th June 1938. One of 18 engines of the class to be rebuilt in 1948, it was withdrawn from Llandudno Junction shed as No. 45525 in May 1963.

Situated three miles to the east of Llandudno, Colwyn (renamed Colwyn Bay in 1876) has become one of the largest towns to the west of the Merseyside region, providing attractive holiday resort facilities on the 18 mile stretch of the North Wales shoreline.

45526 MORECAMBE AND HEYSHAM

Built at Derby Works in March 1933 and later rebuilt in February 1947, this loco was named at Morecambe station by the town's mayor on 6th October 1937, the crest bearing the civic coat-of-arms being added above the nameplate that December. It spent most of its working life at Carlisle Upperby, from where it was withdrawn in October 1964.

The Municipal Borough of Morecambe and Heysham was formed in 1928 by the merger of Morecambe Municipal Borough and Heysham Urban District Council, the two towns being only 3 miles apart. However, this new borough was abolished in 1974 when it was incorporated into the City of Lancaster.

The 1871 takeover by Midland Railway of the troublesome 'Little' North Western Railway (not to be confused with the LNWR) gave the MR a line from Settle Junction to Morecambe, allowing Leeds/Bradford to Morecambe through traffic. Whilst not having quite the pull of Blackpool, so popular was Morecambe with residents of the West Riding textile towns, that it became known as 'Bradford-by-the-Sea'. In fact, its location, set in Morecambe Bay with the mountains of the Lake District at one side was superior to that of its noisy neighbour to the south. One of its main attractions, then and now, is the magnificent classic Art-Deco Grade 2*-listed Midland Hotel, built by the LMS in 1933 across the road from the station. With 44 rooms and a dining area looking out on the Irish Sea, it is worth the trip on its own, although a wonderful statue of Eric further along the impressive promenade is also a 'must see'.

19TH AUGUST 1961 •
No. 45521 RHYL arrives at Halifax with a Summer Saturday excursion for the Lancashire coast.
PHOTO: LARRY FULLWOOD
© THE TRANSPORT TREASURY

No. 45522 PRESTATYN passes Menai Bridge signal box on its way towards Holyhead with a parcels working.
PHOTO: E. NORMAN KNEALE
© THE TRANSPORT TREASURY

11TH MAY 1963 •
No. 45523 BANGOR pictured at Berkhamsted returning Wigan supporters home after their team had been beaten in the Rugby League Cup Final at Wembley Stadium 25–10 by Wakefield Trinity in front of a crowd of 84,492 spectators.
PHOTO: FLINT/HARBART
© THE TRANSPORT TREASURY

No. 45525 COLWYN BAY waiting departure at Berkhamsted station, date not recorded.
PHOTO: DEREK POTTON
© THE TRANSPORT TREASURY

1ST JUNE 1963 • No. 45526 MORECAMBE AND HEYSHAM at Etterby Junction, Carlisle.
PHOTO: ROBERT ANDERSON/ THE TRANSPORT LIBRARY

AUGUST 1964 • No. 45527 SOUTHPORT departs from Glasgow Central with a southbound express.
PHOTO: W. A. C. SMITH
© THE TRANSPORT TREASURY

2ND MAY 1959 • No. 45528 (yet to be named R.E.M.E) awaits departure from Birmingham New Street with the 11.20am to Glasgow Central. PHOTO: ARTHUR W. CUNDALL © THE TRANSPORT TREASURY

45527 SOUTHPORT

This was one of the Derby-built series of 'Patriot' 4-6-0s of 1933, being rebuilt by British Railways in September 1948. No. 5527 was officially named by the LMS in 1937 and carried a crest bearing the town's civic coat-of-arms above the nameplates. A Liverpool Edge Hill-based locomotive, both before and after rebuilding, from December 1933 to March 1961, it was withdrawn from Carlisle Kingmoor depot in December 1964.

A large seaside town in Merseyside, the town was founded in 1792 when William Sutton, as Innkeeper, built a bathing house at what is now Southport's well-known main shopping location, Lord Street. It is also famous for having the second-longest seaside pleasure pier in the British Isles, at the end of which Morrell's Marvellous Marionettes were a star attraction with young children (including my late wife) in the 1950s.

The town was reached by the railway on 24th July 1848 from Liverpool with a line from Manchester following in 1855 to the LYR terminus at Chapel Street. The Cheshire Lines Committee also opened a line to Lord Street station in 1882. A hugely popular destination for holidays and day trips throughout Lancashire and Yorkshire, particularly so for those who wanted something perceived as a little more refined than Blackpool.

45528 R.E.M.E.

The Corps of Royal Electrical and Mechanical Engineers (REME) provides engineering support to maintain and repair the vast array of British Army equipment. It was formed on 1st October 1942 from the technical units of the Royal Engineers, the Royal Army Service Corps, and the engineering branch of the Royal Army Service Corps. They will be found wherever the Army is operating, at home or overseas.

Built at Derby in 1933 and covering nearly 1½ million miles in service, the loco was rebuilt with a taper boiler in August 1947, after which it was continually moved between Camden and Crewe North for 10 years. After three years at Willesden, it ended its days at Annesley shed, working down the ex-Great Central main line following that route being switched from the Eastern Region of BR to the Midland Region. There is an exception to every rule and 45528 was the only rebuilt Patriot to be outshopped without a name, an omission which was rectified, following a staff suggestion, without ceremony, on 2nd October 1959, only three years before withdrawal.

45529 STEPHENSON

Named after 'the Father of Railways', George Stephenson, but not given his Christian name, this loco received its nameplates on 12th August 1948 at Chesterfield Market Place station during a ceremony to honour the centenary of the great man's death. He had lived in the town from 1837 and died there at his home, Tapton House, which is now used as offices by Chesterfield Borough Council. Stephenson's 94 acre park is open for free use by the general public.

After being rebuilt in 1947, this engine was continually transferred between Camden and Crewe North for ten years before moving to Willesden and, finally, as with a few other rebuilds, it ended its days at the former Great Central depot at Annesley before withdrawal in 1964.

45530 SIR FRANK REE

Sir Frank Ree was General Manager of the LNWR and the North London Railway from 1909. Knighted in 1913, he died on 17th February 1914.

The name was first allocated to 5501, until that changed to *St. Dunstan's*, and was transferred to 5530 in 1937. Although this loco was the first of the Patriots to be rebuilt with the 2A taper boiler, it was the final survivor of the whole class, having covered

AUGUST 1948 • A works photo of freshly repainted No. 45529 STEPHENSON. PHOTO: COURTESY, PETER SIKES

a staggering 1,740,000, miles. It lasted until December 1965 – three years after the last original example, 45550.

45531 SIR FREDERICK HARRISON

This name was originally allocated to 45524, which was then renamed *Blackpool*. Lieutenant Colonel Sir Frederick Harrison (1844-1914) was railway manager and an officer in the British Army's Engineer and Railway Volunteer Staff Corps. At the age of 20, he became a clerk on the LNWR at Shrewsbury. He rose through the ranks, working at Euston, Liverpool, and Chester, becoming Chief Goods Manager of the LNWR and finally, from 1893 until 1908, General Manager of the LNWR. He was made a Knight Bachelor in 1902 and later became Deputy Chairman of the South Eastern Railway.

Sir Frederick Harrison was the last of eight to be rebuilt by the LMS, the other ten being so converted in BR days. It was once one of the two class members adorned with an experimental light apple green livery *(see below)*, but this did not wear well and was soon abandoned.

45532 ILLUSTRIOUS

Named after HMS Illustrious, a British battleship of the Majestic Class, which was built at Chatham, launched in 1896, recommissioned at Devonport in 1912 for service with Third Fleet, and scrapped in 1920. She was the third warship of the Royal Navy to bear this name.

Spending time at Crewe, Camden, and Nottingham, this was a name transferred from Claughton 6011 which had been built in 1921 and named in 1923. The Patriot was turned out named from Crewe in April 1933, rebuilt in June 1948, and withdrawn in February 1964.

45534 E TOOTAL BROADHURST

The name was carried by 45525 until 1st June 1937 and transferred to 45534 in September of the same year. Towards the end of its life, it was a regular performer on the North Wales coast line whilst allocated to Llandudno Junction.

Named after Sir Edward Tootal Broadhurst, (19th August 1858 to 2nd February 1922) who was a director and eventually chairman of Tootal Broadhurst Lee, one of the largest cotton manufacturers in Manchester (older readers will remember Tootal ties, once the undoubted market leader in neckwear). He was also the chairman of the Manchester and Liverpool District Bank (which became part of the newly merged National Westminster Bank on 1st January 1970), and a director of the LNWR and the Atlas Insurance Company. He was High Sheriff of Lancashire in 1906/07.

Part of the committee which organised the recruitment of the Manchester Pals battalions in the First World War, Broadhurst donated land in Moston to Manchester Corporation in 1920, as an offering of thanks to the men and women of Manchester for their work in the First World War, now known as Broadhurst Park. Near to Newton Heath, this is now a community facility,

Above: 1st September 1951 • No. 45532 Illustrious departs from Tebay station.
Photo: Neville Stead Collection © The Transport Treasury

Right: June 1949 • Before it acquired smoke deflectors, No. 45534 E. Tootal Broadhurst is pictured at Glasgow Central.
Photo: E. Norman Kneale © The Transport Treasury

Below: 27th July 1963 • No. 45535 Sir Herbert Walker K.C.B. hauls 15 empty coaches past Sanquhar signal box on its way from Corkerhill to Hazel Grove. The stock was being returned south after the Glasgow Fair holiday.
Photo: W. A. C. Smith © The Transport Treasury

used every day by local people. It hosts a range of sporting and recreational activities, hugely appropriately being home to FC United of Manchester, a club set up by disillusioned fans of Manchester United, which itself started life in 1878 as Newton Heath Lancashire & Yorkshire Railway Football Club, by which name it was known until 1902.

Next to the last to be rebuilt, in December 1948, it lasted until May 1964.

45535 SIR HERBERT WALKER K.C.B.

Sir Herbert Ashcombe Walker K.C.B. (15th May 1868 to 29th September 1949) joined the LNWR as a clerk at Euston in April 1885. In 1893 he was made Assistant District Superintendent, North Wales Division and in 1902 appointed District Superintendent at Euston, visiting the US to study American practice. From 1st January 1912, he was appointed General Manager of the London and South Western Railway, where he instigated the programme of third rail electrification. He was knighted in March 1915.

In January 1917, he was acting chairman of the Railway Executive Committee, for which he was made a Knight Commander of the Order of the Bath (KCB). Appointed General Manager of the Southern Railway in 1923, he encouraged the further development of the electrification programme. On retirement in 1937, he became a Director of the Southern Railway until nationalisation in 1947. A commanding presence with a remarkable memory and a strong advocate of a Channel Tunnel, he died on 29th September 1949. There is a memorial to him in the stonework of Waterloo station.

The loco was Derby-built in May 1933, named in 1937, rebuilt in 1948, and withdrawn October 1963.

45536 PRIVATE W. WOOD, V.C.

Wilfred Wood was an engine cleaner at Stockport Edgeley shed when he joined up in 1916 at the age of 19. Joining the Northumberland Fusiliers, he was awarded the Victoria Cross following his actions at the Battle of Vittorio Veneto, Italy in 1918. He charged enemy machine guns on his own, helping cause the surrender of over 300 men. As his citation for the Victoria Cross put it: *"The conspicuous valour and initiative of this gallant soldier in the face of intense rifle and machine-gun fire was beyond all praise"*.

After the war Wilfred returned to the railways, eventually becoming an engine driver and a supervisor. He retired in 1960 and died in 1982. The nameplate which bears his name can be seen in the Northumberland Fusiliers Museum, Alnwick Castle, a memorial to an ordinary man capable of extraordinary feats.

In view of the proximity to Stockport it is appropriate that, apart from 9 months at Bushbury in 1949/50, *Private W. Wood, V.C.* was allocated to Longsight depot after being reboilered in November 1948 until withdrawal on 10th December 1962 with 1,638,862 miles on the clock. Note that the nameplate bore a comma after his name, which is not the case with that for Ernest Sykes.

45540 SIR ROBERT TURNBULL

Sir Robert Turnbull MVO (21st February 1852 – 22nd February 1925) was a British railway manager, joining the LNWR in 1868 and becoming its General Manager in 1914. He was made a member of the Royal Victorian Order in the 1911 New Year's Honours and knighted in 1913. He served as a Lieutenant Colonel in the Engineer and Railway Staff Corps.

The loco was one of the two Patriots repainted in experimental LNER light green and was exhibited in this livery at Marylebone station on 20th December 1948 in company with 4-6-2 No. 46201 *Princess Elizabeth* in LNWR black.

45545 PLANET *(see page 51.)*

WHAT OF THE UNNAMED PATRIOTS?

The following names were selected in 1942, but never used:

- 45505: *Wemyss Bay*
- 45509: *Commando*
- 45513: *Sir W. A. Stanier*
- 45529: *Air Training Corps*
- 45542: *Dunoon*
- 45545: *The Royal Marine*
- 45549: *RAMC*
- 45550: *Sir Henry Fowler*
- 45551: *Rothesay*

Additionally, other names which had been proposed to fit to the rebuilt locos, but never allocated were: *Champion*, *Courier*, *Goliath*, *Harlequin*, *Oregon*, *Velocipede* and *Vulcan*.

BIBLIOGRAPHY

A Detailed History of The Stanier Class Five 4-6-0s Volume 1 – John Jennison, RCTS 2013
British Railways Steam Locomotives 1948-1968 – Hugh Longworth, OPC 2005
British Railways Steam Locomotive Allocations – Hugh Longworth, OPC 2011
The Locomotive Giggleswick – Nigel Mussett, Kirkdale Publications 2003
The Power of the Patriots – J. S. Whiteley and G. W. Morrison, OPC 1997
Names & Nameplates of British Steam Locos No. 1 LMS – Alex Henley, Heyday 1984
LMS Locomotive Names – Rev. John Goodman, RCTS 1994
Claughton & Patriot 4-6-0s – G. Toms and R. J. Essery, Wild Swan 2006
The Patriots a Pictorial Record – Peter Sikes, The LMS Patriot Project 2019
The Warrior – Various LMS-Patriot Company magazines
The Lancashire & Yorkshire Railway Volume 1 – John Marshall, David & Charles 1969

No. 45536 Private W. Wood, V.C. and 'Royal Scot'
No. 46130 West Yorkshire Regiment at Euston.
Photo: Milepost 92½ © The Transport Treasury

An undated shot of No. 45540 Sir Robert Turnbull passing through Golcar station in the Colne Valley between Huddersfield and Stalybridge.
Photo: Neville Stead Collection
© The Transport Treasury

45545 Planet

This engine was turned out from Crewe works on 27th March 1934, being the only one of the last ten numbered Patriots to be rebuilt, which it was on 5th November 1948. Withdrawn in May 1964 it had amassed 1,552,468 miles in service. The nameplates fitted to this loco were those formerly carried by Royal Scot 46131 but had to be shortened to fit. The Scot was subsequently renamed *The Royal Warwickshire Regiment*.

This was a resurrection of the name *Planet*, the ninth locomotive built for the Liverpool & Manchester Railway by Robert Stephenson in 1830 and a development of Stephenson's earlier *Rocket*. *Planet* was the forerunner of a class of nine engines, six built by Stephenson and a further three built by the Leeds company of Murray and Wood. A significant difference between *Rocket* and *Planet* was the use of inside cylinders. This came about after Trevithick told Stephenson that whilst rebuilding Cornish mine engines, he got a substantial improvement on performance and efficiency if he enclosed the cylinders in an insulating jacket. To achieve this Stephenson moved the cylinders from their outside position and placed them in the smokebox to reduce the heat loss through the wall of the cylinder.

17TH JULY 1963 • Shap is the location for rebuilt Patriot No. **45545 PLANET** which is climbing the 1 in 75 gradient just south of Scout Green with a Class C fully fitted freight. Banking assistance has not been taken as the load is well within the capacity of the powerful 4-6-0.
PHOTO: © DAVID P. WILLIAMS COLOUR ARCHIVE/THE TRANSPORT LIBRARY

THE CALEDONIAN RAILWAY CLASS 294 'JUMBO' 0-6-0S
by David Anderson

13TH JANUARY 1962
'Jumbo' 0-6-0 No. 57345 waits patiently with its freight at Coupar Angus.
PHOTO: NORRIS FORREST © THE TRANSPORT TREASURY

In 1883, Dugald Drummond, the locomotive superintendent of the Caledonian Railway produced his first engine design for the company – an 0-6-0 goods locomotive which was to be the prototype for all later types for the Caledonian Railway. The engines had the distinction of being the most numerous class of engine to work on the Scottish railway system.

The '294' 0-6-0s were based on the 18" inside cylinder goods locomotives which Drummond had built for the North British Railway in 1876. Built by Neilson & Company at St. Rollox Works, Glasgow, a total of 224 'Jumbo' 0-6-0s were built during a period of fourteen years under the direction of four successive Caledonian Railway locomotive superintendents – Drummond, Smellie, Lambie and McIntosh. Because of the engines' ability to move heavy freight loads the class became affectionately known as 'Jumbos'. Taken over by the LMSR at the 1923 grouping, and by nationalisation in 1948, only six engines had been scrapped, a tribute to their simple design and reliability.

The 0-6-0s were mainly used on coalfield workings, iron and steel and dockyard duties as well as branch line passenger work within the industrial lowlands of Central Scotland. The 'Jumbos' were to become a valuable and enduring asset from their Caledonian Railway origins into the British Railways era and a familiar part of the steam railway landscape for 75 years north of the border. A total of 123 engines of the class were built to Drummond's design, followed by another 38 by John Lambie between 1892 and 1895 with the last series of 83 under McIntosh's tenure between 1895 and 1897.

The main dimensions of the '294' class were a driving wheel diameter of 5 feet, two inside cylinders of 18" x 25", a tractive effort of 17,901 lbs and a weight in working order of 41 tons 6 cwt.

Most of the early engines had steam brakes only, but all of the McIntosh series were Westinghouse brake fitted to enable them to work on fitted freight and passenger trains, and they were painted in Caledonian Railway blue livery. Five engines of the final series were condenser-fitted for duties on the underground Glasgow Central Railway.

The '294' class were paired with 4 ton coal capacity, 2,500 gallon six-wheel tenders, the original engines with underhung springs, but these were replaced with new types. During their lifetime the 0-6-0s received a great variety of tender types in exchange from other locomotives.

On many of the 'Jumbos' the original chimneys were replaced by a stovepipe variety, a change which did not enhance their appearance, whilst a few engines were given North British Railway pattern chimneys during overhaul at Inverurie Works.

The numbering of the 0-6-0s was complex and confusing and not always in the order of their construction.

In 1917, 25 engines of the class were requisitioned for overseas service with the Railway Operating Department (R.O.D.) and were returned to the Caledonian Railway in 1919.

30TH JULY 1949 • LMS 'Jumbo' 0-6-0 No. 17415 at Motherwell.
PHOTO: JOHN ROBERSTON © THE TRANSPORT TREASURY

On their takeover by the LMSR the class were allocated the numbers 17230–17473 and they were classified as power class 2F, and painted unlined black. Apart from their extensive use from the 1930s on the former Glasgow & South Western Railway at Ayr and Hurlford, the 'Jumbos' spent most of their working lives on the Caledonian Railway, with the main allocations being at Polmadie, Motherwell, Hamilton, Corkerhill and Dawsholm, with a few examples shedded at Stirling and Grangemouth.

In British Railways ownership the 0-6-0s were renumbered 57230–57473 and all were painted in black livery. Apart from their concentrations of activity in the Central Lowlands, the 'Jumbos' were used on branch line work at Stranraer, for the Whithorn line, Dumfries, for the Kirkcudbright branch, Oban and Killin, and Forfar for the Strathmore lines. Duties also included station and yard pilots at Perth and Aberdeen.

Large scale withdrawals of the 0-6-0s took place during the early 1960s and by the end of 1963 only 16 of the class remained in service with the redundant 'Jumbos' being stored at various sheds awaiting disposal. The last four, Nos. 57261, 57296, 57355 and 57375 were not officially withdrawn until November 1963. No. 57375 of Stranrear shed was used on enthusiasts' steam excursions in its final days.

In summary, the Dugald Drummond 0-6-0 goods engines of 1883-1897 were a simple, robust and reliable design which remained virtually unaltered and still met the needs of Scottish freight duties in the final days of steam working.

9TH SEPTEMBER 1955 • British Railways 'Jumbo' Class 2F 0-6-0 No. 57451 at Carstairs Junction. PHOTO: DAVID ANDERSON © THE TRANSPORT TREASURY

CALEDONIAN RAILWAY 294 CLASS TWO-CYLINDER 0-6-0					
Leading Dimensions	A	B	C	D	E
Cylinders (inch)	18 x 26	18 x 26	18 x 26	18 x 26	18 x 26
Coupled Wheels (feet/inch)	5.0	5.0	5.0	5.0	5.0
Boiler:					
Pressure	150 lbs	150 lbs	150 lbs	150 lbs	180 lbs[u]
Diameter	4ft 5⅛ins	4ft 5⅛ins	4ft 5⅛ins	4ft 6¼ins	4ft 6¼ins
Length	10ft 3⅛ins	10ft 3⅛ins	10ft 3⅛ins	10ft 3⅛ins	10ft 3⅛ins
Height of Centre	7ft 3ins	7ft 3ins	7ft 3ins	7ft 3ins	7ft 3ins
Tractive effort at 85% of boiler pressure	17,901 lbs	17,901 lbs	17,901 lbs	17,901 lbs	21,481 lbs[u]
Heating Surface:					
Tubes	1095.6 sq.ft.	1089.68 sq.ft.	1071.5 sq.ft.	1056.8 sq.ft.	1056.8 sq.ft.
Firebox	113.0 sq.ft.	112.62 sq.ft.	112.62 sq.ft.	112.4 sq.ft.	112.4 sq.ft.
Total	1208.6 sq.ft.	1202.3 sq.ft.	1184.12 sq.ft.	1169.2 sq.ft.	1169.2 sq.ft.
Grate Area	19.5 sq.ft.	19.5 sq.ft.	19.5 sq.ft.	19.5 sq.ft.	19.5 sq.ft.
Tender:					
Water	2,500 gallons[v]	2,500 gallons[w]	2,500 gallons[x]	2,840 gallons[y]	2,800 gallons[z]
Coal	4½ tons	4½ tons	4½ tons	4½ tons	4½ tons
Wheelbase:					
Engine	16ft 3ins	16ft 3ins	16ft 3ins	16ft 3ins	16ft 3ins
Tender	13ft 0ins	13ft 0ins	13ft 0ins	13ft 0ins	13ft 0ins
Total	37ft 4½ins	37ft 4½ins	37ft 4½ins	37ft 4½ins	37ft 4½ins
Length over Buffers	49ft 10¾ins	49ft 10¾ins	49ft 10¾ins	49ft 10¾ins	49ft 10¾ins
Weight:					
Engine	41 tons 6 cwt	41 tons 6 cwt	40 tons 6 cwt	41 tons 6 cwt[y]	42 tons 4 cwt
Tender	34 tons 10 cwt	34 tons 10 cwt	35 tons 16 cwt[x]	33 tons 19¾ cwt[y]	33 tons 19¾ cwt[z]
Total	75 tons 16 cwt	75 tons 16 cwt	76 tons 2 cwt	75 tons 5¾ cwt	76 tons 3¾ cwt

A   Engines designed by D. Drummond
B   Engines designed by H. Smellie
C   Engines designed by J. Lambie
D   Engines designed by J. F. McIntosh
E   Engines fitted with boiler by LMSR
u   LMSR boiler built to carry a pressure of 180lbs per sq.in. but actual working pressure varied between 150 and 180lbs per sq.in. according to state of frames and tractive effort varied between 17,-901 and 21,481lbs per sq.in.
v   Tender capacity varied between 2,500 and 2,840 gallons and weight between 34 tons 10 cwt and 35 tons 16 cwt.
w   Only tender used.
x   Later tenders carried 2,800 gallons, early tenders weighed 34 tons 10 cwt.
y   Engines fitted for condensing carried only 2,800 gallons of water, engine weight 42 tons 4 cwt and tender weight 34 tones 17 cwt.
z   Tenders varied according to their origin.

ABOVE: Caledonian Railway No. 299 in original condition as built by Neilson & Co. in 1883, with a breakdown train at an unrecorded location.

RIGHT: 1883-built Caledonian Railway No. 337 at St. Rollox.

BELOW: Blue-liveried No. 748 at Perth, built at St. Rollox in 1896 and fitted with a Westinghouse pump. It became LMSR No. 17415 and was withdrawn in 1949.

ALL PHOTOS: DAVID ANDERSON COLLECTION

ABOVE: LMS No. 17424 (formerly CR No. 757) pictured at Callander station. This loco made it into British Railways service as No. 57424 and was withdrawn in 1959.

BELOW: LMS No. 17230 (formerly CR No. 299, see opposite page, top photo) approaches Dumfries with a freight working from Stranraer. Withdrawn by BR as No. 57230 in 1956.

2ND JUNE 1956 • Drummond Caledonian Class 294 'Jumbo' Standard Goods 2F 0-6-0 No. 57345 at Perth MPD.
PHOTO: DAVID ANDERSON © THE TRANSPORT TREASURY

29TH SEPTEMBER 1956 • Drummond Caledonian Class 294 'Jumbo' Standard Goods 2F 0-6-0 No. 57287 pictured between Gorgie and Craiglockhart on an empty mineral train made up of predominantly wooden stock.
PHOTO: DAVID ANDERSON © THE TRANSPORT TREASURY

17th August 1963 • 'Jumbo' 0-6-0 No. 57261 on shunting duties at Gleneagles station. Photo: Neville Stead Collection © The Transport Treasury

21st July 1956 • 'Jumbo' 0-6-0 No. 57386 takes a break from shunting duties at Carstairs No. 1 signal box. Photo: W. A. C. Smith © The Transport Treasury

THE MIDLAND 'SINGLES'
BY DAVID CULLEN

SEPTEMBER 1961 • Johnson Midland Railway 'Spinner' No. 118 outside Derby Roundhouse.
PHOTO: NEIL DAVENPORT, ONLINE TRANSPORT/THE TRANSPORT LIBRARY

Whereas most steam locomotives ran on multiple sets of driving wheels, in the early decades of the nineteenth century many were equipped with just a single pair. Bogies were also fitted for weight bearing and stability, frequently, though not always, both ahead and rearward. Most single drivers were larger than those on coupled locomotives, many considerably so. On these engines, superstructures above would greatly complement the arrangement, giving an impressive, often artistic appearance.

The arrangements were as follows: 0-2-2 in which the locomotive had no leading bogies with two driving wheels, plus two trailing bogies. This was the arrangement on Stephensons' *Rocket*, winner of the Rainhill Locomotive Trials of October 1829. *Northumbrian*, also an 0-2-2, hauled the first train on the Liverpool and Manchester Railway in September 1830. Appearing in the same year, the Liverpool & Manchester Railway Planet class were 2-2-0s. The 2-2-2 notation was adopted during the mid-19th century by the Great Western Railway. 4-2-4 was applied to the Bristol & Exeter Railway Pearson Singles of 1854, tank locomotives having colossal 8 foot 10 inch driving wheels. Then there was the 4-2-2.

It is the 4-2-2 notation that is relevant here, famously fitted to the Great Northern Railway 'Stirling Singles' of 1870, locomotives described as engineering artistry. In those days of wooden, non-corridor carriages, these engines coped more than adequately, considering how providing crucial rail grip was down to that one pair of driving wheels. 'Stirling Singles' would handle many top passenger expresses, including the legendary, at first unofficially named long-distance express, *The Flying Scotsman*.

Moving on seventeen years from the Stirling Singles, the Midland Railway introduced their 4-2-2 Singles, designed by Chief Mechanical Engineer, Mr. Samuel W. Johnson. Five sub-classes were constructed at the Midland's Derby Works, their number totalling ninety-five over a period of thirteen years.

While 'Singles', including the Midland and GNR's, operated for many more years, their construction generally ceased in the late 19th century. The only railway to continue building them was the Midland, albeit only into 1900.

The railway operated frequent, fast trains of modest weight, meaning these services and the new 'Singles' were perfectly matched. Their introduction was however a retrograde move, following twenty-one years of constructing more up-to-date locomotives having additional drivers coupled. In addition, the Midland operated a 'small engine' policy which was never subject to revision. Their effectiveness was further sustained following Derby Works manager Mr. Francis Holt's invention of steam operated gear for sanding slippery rails, ensuring superior adhesion for hauling. The development would soon replace less effective gravity-fed sanding.

Driving wheel diameter regulates rail distance covered by each revolution. So, a further benefit of the 'Singles' was having large driving wheels meant fewer rotations required for route coverage. This resulted in reduction of motion for pistons, valves, valve gear and connecting rods, advantageous in an era prior to lubrication advancement.

There were five sub-classes, between them carrying a number of operational variations. These were the 25 class, eighteen in number built 1887-90; the 1853s of 1889-93 numbering forty-two; ten 179s of 1893-96; fifteen 115s of 1896-99 constructed in two batches; plus ten 2601s built 1899-1900. Giving locomotives names was not normal Midland practice, however, the exception came with one of the final batch. This engine, No. 2601, was honoured with the epithet *Princess of Wales*. It participated in the Paris Exposition in 1900 during which it received *The Grand Prix*.

Technical Data

All ninety-five locomotives operated with two inside cylinders fitted out of sight between the main frames. Standard practice during the nineteenth century, inside cylinders had the advantage of producing less nosing oscillation than ones fitted outside. Inside units however, were by then largely on borrowed time. The valves were controlled by Stephenson link gear, again located between the frames.

To the variations. The 25 class locomotives had 7 foot 4 inch driving wheels, cylinders of 18 inches diameter by 26 inch stroke, 'slide' type valves and a maximum boiler pressure of 160 lbs per sq.in. The class 1853s had 7 foot 6 inch driving wheels, cylinders of 18½ x 26 inch and again 'slide' valves. The Class 179s had 19 x 26 inch cylinders, 'piston' type valves and 7 foot 6 inch drivers. Maximum steam pressure was 160 lbs lbs per sq.in. The next sub-class, the 115s, will be covered forthwith. Finally, the 2601 class had drivers of 7 foot 9½ inch diameter, cylinders of 19½ x 26 inch, 'piston' type valves and maximum steam pressure of 180 lbs per sq.in.

While the 1853s were the most numerous, the 115 class received the greatest admiration. Indeed, their design became recognised as among the most impressive of all British locomotives. Their dark red livery known as 'Crimson Lake' was highly popular with the Midland Railway and its elegance a major source of pride. Subsequently all the engines were kept spotlessly clean. Further, it is said that meticulous inspections were carried out to ensure even out-of-sight areas including back surfaces of wheels were receiving attention. In addition, drip

Two unidentified ex-MR locos – a 4-2-2 'Spinner' piloting a Belpair-boilered 4-4-0 – at Milepost 10, which is likely to be south of Bedford on the Midland Main Line.
PHOTO: MILEPOST 92½ © THE TRANSPORT TREASURY

trays were placed beneath them on depot to avoid oil defiling the immaculately kept floor. Such rigorousness being the norm, as would be expected their mechanical conditions were maintained to the highest standards.

The 115s' cylinders were of 19½ ins diameter by 26 ins stroke, their lateral centre lines 2 foot 3 inches apart. They were set at an incline of 1 in 16 above the horizontal. Located beneath, inclined downwards to the same degree, their 'piston' type valves were 8 inches diameter, of 2½ inches length and had a maximum travel of $3^{11}/_{16}$ inches. They were worked by two inside sets of Stephenson link motions controlled manually by cab screw reversers. Live steam pipes were formed from copper and 4 inches in diameter. The blastpipe nozzles were 5 inches.

Driving wheel diameter was an impressive 7 foot 9½ inches. This earned the locomotives the nickname '*Spinners*' from the rapid whirling of the wheels when running at speed. Leading bogies were of 3 foot 9½ inches, trailing bogies 4 foot 4 inches and the six tender wheels 4 foot 2½ inches. The tenders were fitted with handbrakes. All wheels were liveried to match the Crimson Lake superstructures above. Overall length between front and rear buffers was 52 foot 7½ inches. Height from rail level to chimney crown was 13 foot 1½ inches and to boiler mid-line 7 foot 10 inches.

Boilers were of the Midland 'E' type containing 236 x 1⅝ inches diameter copper fire tubes giving a combined surface area of

JULY 1957 • The cab of No. 118, illustrating the lack of protection from the elements. The locomotive was being positioned for display at Leicester Belgrave Road station.
PHOTO: H. CARTWRIGHT © THE TRANSPORT TREASURY

1,105 sq.ft. Together with 128 sq.ft copper fireboxes, 1,233 sq.ft of evaporative surface was provided. Distance between front and rear tubeplates was a fraction over 11 feet. Boiler barrel thickness was half an inch. Maximum boiler pressure was 170 lbs per sq.in.

The copper firebox interiors were a fraction under 6 foot 4 inches long with a width of 3 foot 4½ inches. Grate areas were 21.3 sq.ft. Smokeboxes were constructed of steel, 2 foot 11 inches in length with internal diameters of 4 feet. In contrast with the main livery, they were coloured black.

At this stage of locomotive development there was no superheating, the equipment not being developed for around another two decades. Locomotive tops were adorned with tall black chimneys matching the smokeboxes, liveried steam domes and brass safety valve shroudings.

Weights in full working order were 47 tons 2 cwt for the locomotives and 38 tons 7 cwt for the tenders, a total of 85 tons 9 cwt. Maximum axle load was 18½ tons, as was adhesive weight, this obviously created by the single drivers. The front bogies carried 15 tons 16 cwt and the trailing 12 tons 15 cwt. Tenders carried coal and water stocks of 4½ tons and mostly 3,500 gallons, although some engines were paired with tenders carrying 2,950 gallons.

Tractive effort, a value indicating haulage capability, was 15,279 lbs. For a simple expansion steam locomotive this is produced using the following formula in which all measurements; apart from steam pressure obviously, are expressed in inches.

Beginning with 'squaring' the cylinder diameter (multiplying this by itself) the result is then multiplied by the piston stroke. The result is multiplied by a reckoned working steam pressure of 85% of maximum, allowing for a 15% drop in flowing from boiler to cylinders. For these locomotives this was 144½ lbs per sq.in. The result so far is finally divided by the driving wheel diameter.

Regarding locomotive numbering. The two batches of 115s were initially allocated 115-119, 120, 121, 123-128, 130 and 131. In 1907 they were renumbered 670-674 and 675-684. Following the 1923 Grouping Programme they retained these numbers under the London, Midland & Scottish Railway.

Back in the day, Midland Railway passenger trains comprised seven or eight bogie carriages totalling 200-250 tons for which the *Spinners* were perfectly suited. In addition, under dry weather conditions affording good rail grip, loads of up to 350 tons were manageable. All these workings would be handled within timetable requirements due to respectable speed capabilities, rates of around 90 mph being recorded.

23RD SEPTEMBER 1961 • Johnson 'Spinner' No. 118 in Derby Works yard.
PHOTO: R. C. RILEY © THE TRANSPORT TREASURY

6TH JULY 1957 • 1896-built Johnson MR Class 115 'Spinner' 4-2-2 at Leicester Belgrave Road (GN). PHOTO: HORACE GAMBLE © THE TRANSPORT TREASURY

The Midland 'small engine' policy ensured the 'Singles' service era was a long one, many continuing operation for up to 40 years. Not only did they give good account of themselves hauling regular passenger expresses, they were perfectly suited to double heading for handling heavier trains. This was an essential and standard procedure for the Midland due to that 'small engine' policy. They made ideal companions for the equally respected Johnson/Deeley 4-4-0 Compounds.

There was however one drawback with the 'Singles' which I am not the first to acknowledge. The cabs were widely exposed to the backs and sides, providing little protection from the elements. Upper sections of the side panels comprised deep, curved cut-outs which would have assisted leaning out for effective forward vision.

Under 3 foot in length, the roofs gave scant cover. A sizeable gap separated the cabs and tenders. Operational controls would have further limited shelter space.

Journeys made in bitter winter weather, blizzards and torrential rain cannot have been pleasant. However, the format was more or less standard then, and I feel that regardless, footplate crews would have felt honoured to operate these classic locomotives.

The final development of the M.R. 'Singles' appeared in 1899 with the 2601s. One visible variation was the steam dome sited over the driving wheels rather than further forward as on the 115s. They were equipped with M.R. 'F' type boilers producing a maximum pressure of 180 lbs per sq.in. Having a comparatively short life, these were withdrawn from 1919-1922, meaning that no class 2601s made it into L.M.S. ownership, the Grouping programme not being introduced until the following year.

With train weights increasing due to larger, heavier carriages and lengths due to increasing passenger requirements, withdrawals of all the 'Singles' were made between 1921 and 1928. Their replacements were generally 4-4-0s. Construction of larger, more powerful 'Singles' was not an option as the height increase required for accommodating larger boilers, cranked axles and connecting rods would have led to instability. Particularly at express speeds.

Twelve 115s survived into L.M.S. days: Nos. 670-678, 680, 682 and 683. By 1927 only three remained. Just one example was preserved following withdrawal in 1928. Originally Midland Railway No. 118, it subsequently became No. 673 and retained the number under the L.M.S. Now a member of the National Collection, it bears the honour of being the oldest preserved British steam locomotive equipped with piston valves during its construction.

INFORMATION SOURCES

The Great Book of Trains by Brian Hollingsworth and Arthur Cook.
Classic British Steam Locomotives by Peter Herring.
Model Steam Locomotives by Henry Greenly.
www.locomotive.fandom.com
www.modelengineeringwebsite.com
www.preservedbritishsteamlocomotives.com
www.railwaywondersoftheworld.com
www.steamindex.com

26TH MAY 1980 • Johnson 'Spinner' 4-2-2 in its LMS livery and numbered 673 parades through the watching crowd at the Rainhill 'Rocket 150' Cavalcade. PHOTO: TREVOR DAVIS © THE TRANSPORT TREASURY

THE RAILWAYS OF LEICESTER

Leicester was one of the first cities (though then a town) to be served by a railway when the Leicester and Swannington Railway built its terminus station at West Bridge on the western side of Leicester in 1832. The L&SR Railway was built to supply the city of Leicester with coal from the mines in the north west of the county, this early entrant to the country's network was later absorbed by the Midland Railway.

At one time Leicester possessed seven railway stations. In addition to the only surviving Leicester station – the Midland Railway's London Road – three other main railway stations existed. The previously mentioned West Bridge, which closed to passengers in 1928, was itself a replacement for the original West Bridge station that opened with the railway in 1832 but replaced by a larger station in 1893. Leicester Belgrave Road (Great Northern Railway) officially opened in 1883 and closed to passengers in 1962 and Leicester Central, part of the final major railway line to be built in the country, the Great Central Railway, opened in 1899, being closed in May 1969. In addition, there were smaller stations within the city boundary at Humberstone Road on the Midland Railway main line, Humberstone on the GNR, and from 1874 until 1918, a halt at Welford Road which comprised a single platform on the Down main line which allowed access to the Cattle Market. At this halt, passengers were allowed to leave the trains but not to board them, necessitating a walk to nearby London Road station to make their return journey.

The following photos show the Midland interest in Leicester with views of West Bridge station and goods yard followed by images south of Leicester on the Midland Main Line through London Road station and then past the shed (15C) to the north.

For an in-depth look at *'The Railways of Leicester'* a new book by Midland Times editor, Peter Sikes, has been published by Transport Treasury Publishing. The book illustrates Leicester's stations and surrounding areas from the extensive catalogue of images from The Transport Library, these are accompanied by a series of detailed track plans *(see page 69 for details)*.

14TH JUNE 1963 • Leicester West Bridge station, which closed to passengers in 1928. PHOTO: R. C. RILEY © THE TRANSPORT TREASURY

14TH JUNE 1963 • Leicester West Bridge station, the row of terraced houses are on Tudor Road. PHOTO: R. C. RILEY © THE TRANSPORT TREASURY

4TH MAY 1962 • 1875-built ex-MR Johnson Class 2F 0-6-0 No. 58143 shunts Leicester West Bridge yard. The loco was based at Coalville (15E). PHOTO: R. C. RILEY © THE TRANSPORT TREASURY

16TH JULY 1965 • BR Standard Class 2MT 2-6-0 No. 78028 on shunting duty at West Bridge yard. The loco has a cut down cab to enable passage through Glenfield Tunnel. PHOTO: HORACE GAMBLE © THE TRANSPORT TREASURY

Ex-LMS Stanier 8F 2-8-0 No. 48128 passes under Welford Road bridge and is about to pass classmate No. 48473 at Cattle Market Sidings, Leicester.
Photo: H. Cartwright © The Transport Treasury

12th June 1959 • Johnson (rebuilt by Fowler) Class 2P 4-4-0 No. 40452 emerges from Knighton Tunnel while working the 5.50pm Leicester-Birmingham service.
Photo: Horace Gamble © The Transport Treasury

L.C.G.B. The North Countryman Rail Tour

The photos on the right show Burton-based Stanier Jubilee Class 6P 4-6-0 45721 IMPREGNABLE travelling through Leicester on the first leg of THE NORTH COUNTRYMAN which took place on 6th June 1964. The Jubilee would travel as far as Whitehall Junction, Leeds where Gresley V2 No. 60923 would take over. The tour then headed to Carlisle via Skipton and Ais Gill before heading back south to Leeds City via Settle Junction. The third and final leg was hauled by Gresley A3 No. 60051 BLINK BONNY to Kings Cross via Normanton and Doncaster, booked to arrive in the capital at 22.12, an amazing 606½ miles later. The booked timings for the first leg are shown below.

RIGHT (FROM TOP): The train passes Leicester Cattle Market Sidings signal box then enters the cutting south of London Road station before passing through platform 2 and under Swain Street bridge with Leicester North signal box visible to the right of the train.
PHOTOS: DAVID MOORE © THE TRANSPORT TREASURY

1X46 TIMINGS – SATURDAY 6TH JUNE 1964

Miles/Chains	Locwation	Booked Timings
0	London St. Pancras	08.10d
3.44	Finchley Road	08.17½
6.78	Hendon	08.22
19.71	St. Albans City	08.35
30.20	Luton Midland Road	08.44½
49.65	Bedford Midland Road	09.00½
59.51	Sharnbrook Summit	09.10
65.05	Wellingborough Midland Rd	09.18½
72.02	Kettering	09.25
78.39	Desborough North	09.33½
82.75	Market Harborough	09.38
89.60	Kibworth North	09.44½
95.72	Wigston North Junction	09.50½
99.06	Leicester London Road	09.55a – 09.57d
103.63	Syston	10.04
111.46	Loughborough Midland	10.11
119.65	Trent	10.23
125.14	Trowell	10.33
133.38	Pye Bridge	10.51
139.04	Morton Sidings	11.02
142.19	Clay Cross	11.07
146.20	Chesterfield Midland	11.12
151.44	Dronfield	11.21
154.20	Dore & Totley	11.25
158.41	Sheffield Midland	11w32a – 11w38d
161.60	Wincobank Station Junction	11.44
165.77	Chapeltown South	11.52
172.77	Monk Spring Junction	12.03
175.08	Cudworth	12.10
185.11	Normanton	12.27
186.00	Altofts Junction	12.29
195.25	Engine Shed Junction	12.42
195.55	Whitehall Junction, Leeds	12L44a – 12L54d

Timetable reproduced from www.sixbellsjunction.co.uk

ABOVE: 1965 • Standard 9F 2-10-0 No. 92018 occupies the turntable in the now roofless Leicester Roundhouse. Built in 1945, it serviced its last steam locomotive in 1966. After that it was used to store locomotives that were earmarked for a proposed museum in Leicester that sadly never materialsed. The roundhouse was demolished in 1970.
PHOTO: MILEPOST 92½ © THE TRANSPORT TREASURY

LEFT: APRIL 1951 • Ex- Midland Railway Johnson Class IP 0-4-4T No. 58073, on station pilot duties, takes a breather in the carriage sidings at Leicester London Road. PHOTO: ALEC FORD © THE TRANSPORT TREASURY

THE RAILWAYS OF
LEICESTER

FEATURING: WEST BRIDGE, BELGRAVE ROAD, LONDON ROAD AND CENTRAL STATIONS, PLUS SURROUNDING AREAS
112 PAGES, OVER 190 MONOCHROME IMAGES, INCLUDES 12 AREA MAPS

Available from www.ttpublishing.co.uk
£14.95 PLUS P&P

ISBN: 978-1-917776-27-1

OUT NOW

HEST BANK

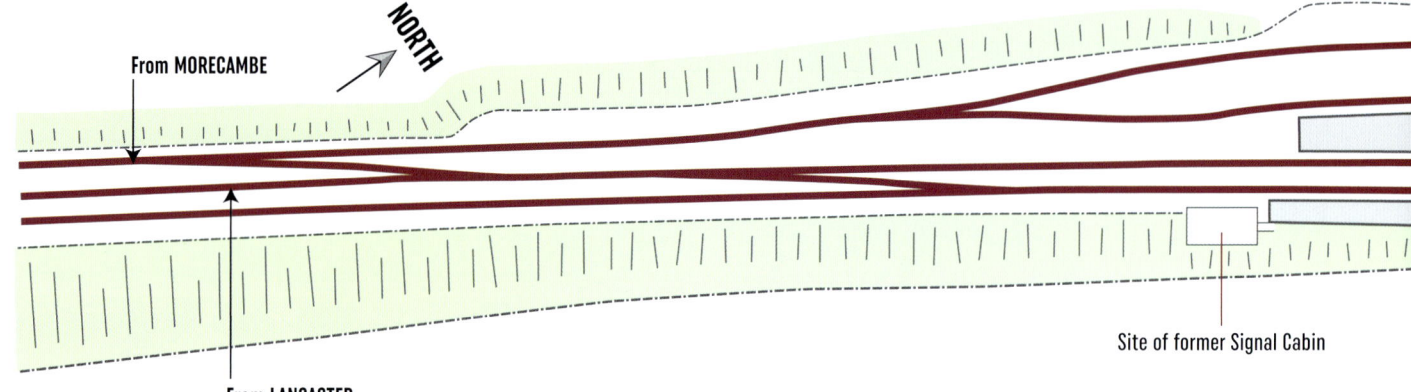

Hest Bank railway station was opened by the Lancaster and Carlisle Railway (L&CR) three miles north of Lancaster Castle station. It was the only station in sight of, and in close proximity, to the coast on the entire length of the West Coast Main Line (WCML). The creation of the WCML was constructed in a piecemeal fashion. By 1841 the London & Birmingham, Grand Junction, North Union amongst others provided a rail route from Euston to Fleetwood, where passengers embarked on a coastal steamer to Ardrossan for Glasgow.

In 1841 a Royal Commission advocated the West Coast as the one route to Scotland, and planning commenced to progress north from Lancaster to Carlisle. The L&CR was authorised on 6th June 1844 and opened to Oxenholme on 23rd September 1846, and to Carlisle in December of the same year. A station at Hest Bank had been proposed in 1845 and was duly provided. It opened in 1846 along with the line. The station continued to serve the village of Hest Bank until its closure in 1969.

Meanwhile, the 'Liittle North Western Railway' had been formed to link Skipton to Lancaster and with the Morecambe Bay & Harbour Railway (renamed Morecambe Harbour & Railway). One branch was to run approximately north-east from Morecambe to join the L&CR at Hest Bank, but this plan was later abandoned. In the 1850s the L&CR had hoped to develop an export trade in coke and other minerals and applied on its own account to build the branch from Hest Bank to Morecambe, permission was received on 13th August 1859. A month after this date the L&CR accepted a working lease by the London and

CIRCA 1900 • **A view of the Lancaster & Carlisle Railway station at Hest Bank.**
PHOTO: LENS OF SUTTON ASSOCIATION/THE TRANSPORT LIBRARY

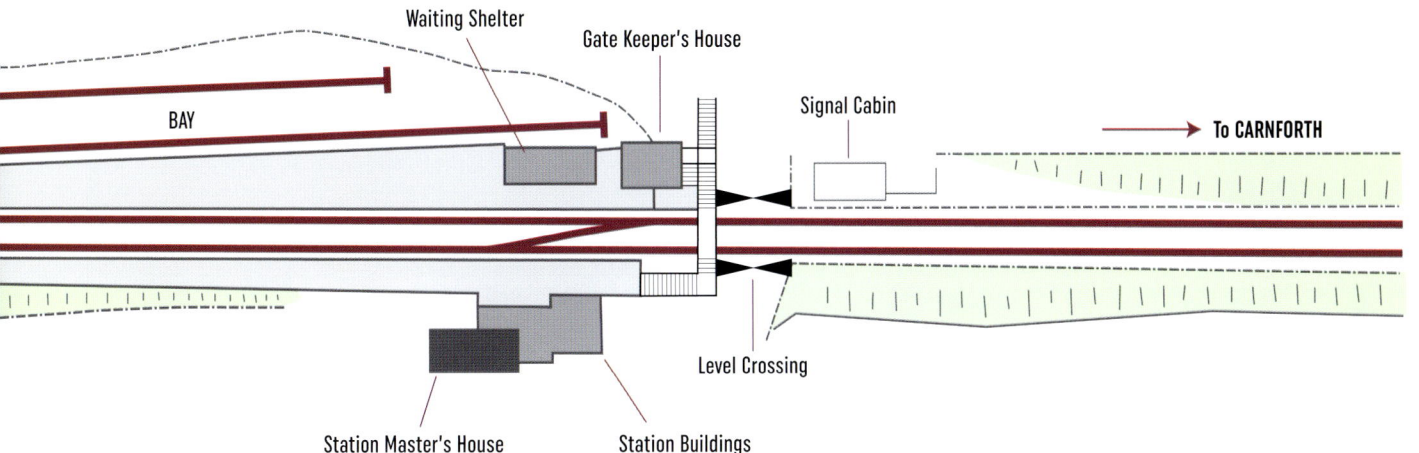

North Western Railway (LNWR). The branch was therefore built by the LNWR and opened as a double track route on 8th August 1864. The hoped-for mineral traffic did not develop and the branch was singled, but with an increase in holiday trade a south facing curve was added to the junction during 1888. The majority of LNWR passenger trains took this route and beyond Bare Lane, where the old and new lines joined, was redoubled and is still open as the Morecambe Branch Line. The section from Bare Lane to Hest Bank remained single.

Hest Bank station itself was of rugged stone construction, a two-storey station house with a booking hall below on the up (southbound) platform. It was next to a level crossing over a minor road linking the A589 with houses, a water treatment works and a caravan park further along the foreshore. A small cottage for the crossing keeper was provided on the down (coastal-facing) side along with a footbridge to connect the platforms. A bay platform and a small goods yard, accessed from the Morecambe branch was provided on the Down side of the line, this closed on 2nd December 1963. However, the track remained and four camping coaches were positioned here by the London Midland Region from 1960 to 1963, then it increased to five until the end of 1969, despite the station finally closing to passengers in February 1969.

There are no traces of the platforms and buildings have disappeared, although the crossing keeper's cottage survived until 2012, as the incoming Morecambe branch was extended along the length of one of the former platforms as part of the layout changes associated with the 1973 WCML electrification scheme.

16TH OCTOBER 1951 • Ex-LMS Stanier 8P Coronation 4-6-2 No. 46257 CITY OF SALFORD passes through Hest Bank on a Perth–Euston express passenger service.
PHOTO: CUMBRIAN RAILWAYS ASSOCIATION/ THE TRANSPORT LIBRARY

JULY 1951 • Ex-LMS 2P 4-4-0 No. 40673 travels south from Pasture Lane bridge double-heading with an unidentified Stanier Class 6P 'Jubilee' 4-6-0 on an Up express passenger at Hest Bank.
PHOTO: MIKE BOLTON/ CUMBRIAN RAILWAYS ASSOCIATION/ THE TRANSPORT LIBRARY

EARLY TO MID-1950S • Fowler 3P 2-6-2T No. 40041 passes through the station at Hest Bank with an Up Class K trip goods.
PHOTO: MIKE BOLTON/CUMBRIAN RAILWAYS ASSOCIATION/THE TRANSPORT LIBRARY

16TH OCTOBER 1951 • Ex-LMS Fowler parallel-boiler 'Royal Scot' 4-6-0 No. 46148 THE MANCHESTER REGIMENT approaches the station at Hest Bank with a Carlisle/Windermere–Euston express. PHOTO: CUMBRIAN RAILWAYS ASSOCIATION/THE TRANSPORT LIBRARY

16TH OCTOBER 1951 • In contrast to the photo above we now see a Stanier Rebuilt 'Royal Scot' Class 7P 4-6-0 in the shape of No. 46104 SCOTTISH BORDERER passing through Hest Bank station. The train is a Manchester and Liverpool to Glasgow and Edinburgh express. PHOTO: CUMBRIAN RAILWAYS ASSOCIATION/THE TRANSPORT LIBRARY

MAY 1954 • Stanier 8P Coronation Class 4-6-2 No. 46242 CITY OF GLASGOW crosses from the Down back to the Up line at Hest Bank after wrong line working with the Up Royal Scot. PHOTO: CUMBRIAN RAILWAYS ASSOCIATION/THE TRANSPORT LIBRARY

19TH JULY 1967 • Stanier 8F No. 48036 passes through Hest Bank station on a short freight. The camping coaches mentioned in the text are seen to the right afford a great view of the coastline. Note the replacement platform edging on the Up platform. This was as a result of an accident at the station in the early hours of 20th May 1965 when a sleeper train was partially derailed by a broken rail just north of the station. Thankfully there were no fatalities but eleven passengers suffered minor injuries. PHOTO: BERNARD MILLS/THE TRANSPORT LIBRARY

19TH JULY 1967 • Stanier Black Five No. 45221 on a weedkiller train passes the Bare Lane/Morecambe spur to the south of Hest Bank station. PHOTO: BERNARD MILLS/THE TRANSPORT LIBRARY

19TH JULY 1967 • Work-stained Stanier 8F 2-8-0 No. 48631 passes the Morecambe spur to the south of Hest Bank station with the 7L00 Oakleigh-Corkickle freight working formed of covered hoppers. PHOTO: BERNARD MILLS/THE TRANSPORT LIBRARY

19TH JULY 1967 • Stanier Black Five No. 44983 works the 3M00 Leeds–Heysham vans through Hest Bank on the Morecambe line.
PHOTO: BERNARD MILLS/THE TRANSPORT LIBRARY

19TH JULY 1967 • The tide is out as Stanier Black Five 4-6-0 No. 45209 reverses light engine towards Hest Bank station on the Morecambe spur. PHOTO: BERNARD MILLS/THE TRANSPORT LIBRARY

19TH JULY 1967 • The signalman is on the 'blower' as 'Britannia' Class 7P 4-6-2 No. 70029 SHOOTING STAR approaches Hest Bank after passing over the water troughs, with a short parcels working. The signal box pictured opened in 1958, replacing the LNWR box to the south of the station, the image gives a view of the lever frame and wheel for operating the level crossing. PHOTO: BERNARD MILLS/THE TRANSPORT LIBRARY

THE R.C. RILEY COLOUR COLLECTION

COMING SOON

 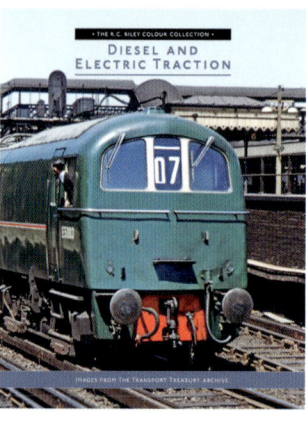 

Transport Treasury Publishing are proud to present a unique series of 15 colour albums featuring the best of the R. C. Riley colour archive. To be released at intervals and printed in strictly limited numbers, the series will grow into a unique record both of the contemporary railway scene and also that of the work of one of Britain's leading transport photographers.

www.ttpublishing.co.uk

PUBLISHED BY TRANSPORT TREASURY PUBLISHING LTD.

THE PLATFORM END

22ND MARCH 1962 • Stanier Class 5MT 4-6-0 No. 44694 departs from Bradford Exchange.
PHOTO: ROBERT ANDERSON/THE TRANSPORT LIBRARY

In future issues our aim is to bring you many differing articles about the LMS, its constituent companies and the London Midland Region of British Railways. We hope to have gone some way to achieving that in this issue.

Midland Times welcomes constructive comment from readers either by way of additional information on subjects already published or suggestions for new topics that you would like to see addressed. The size and diversity of the LMS, due to it being comprised of many different companies, each with their differing ways of operating, shows the complexity of the subject and we will endeavour to be as accurate as possible but would appreciate any comments to the contrary.

We want to use these final pages as your platform for comment and discussion, so please feel free to send your comments to: midlandtimes1884@gmail.com or write to:
Midland Times, Transport Treasury Publishing Ltd.,
16 Highworth Close, High Wycombe HP13 7PJ.

THE PLATFORM END

Highland Railway PO Sorting Van

With reference to Midland Times, No. 8, July '25, page 28, upper left photograph: I would suggest that the bogie carriage behind the horse box is a former Highland Railway Post Office Sorting Van, with its off-side to the lineside clearly visible.

Peter Tatlow's authoritative work, 'Highland Railway Carriages and Wagons', pages 120-124 (Noodle Books, 2014) provides an account of these vehicles, along with black and white photographs and superb line drawings. There is also a very clear view of the postal pick-up net on page 123 of the book.

Similarly, it is fascinating to see that on page 124, that by 1960 the lettering 'Royal Mail' had been applied to the right-hand end of the off-side of the postal van.

Regards, Arnold Tortorella

Shortly before going to press on this edition of Midland Times, Ian Lamb sent in this press release regarding the opening of the replacement station at Forres in 2017. Although we rarely print modern day railway images I thought it would be interesting to compare what was, and the infrastructure as it is now. Ian's Forres Remembered article appears on page 4. (Ed)

First scheduled train from Inverness into the new Forres Station – 17th October 2017

On 8th August 1858, the section of forty miles of railway between Nairn and Keith – part of the Inverness & Aberdeen Junction Railway – was completed. This gave a continuous line from Inverness to Aberdeen and, after passengers changed there, to the south. Thus were the Highlands first joined by rail with the rest of Scotland and Great Britain.

Almost 160 years later, on the site of the original Highland Railway station, a new structure – fit for the 21st Century – was established.

Not even Hurricane Ophelia on 17th October 2017 could put a damper on this inaugural train service, in the pouring rain and high winds, the first scheduled train over the new line arrived at the rebuilt Forres station.

Regards, Ian Lamb

16th June 1962 • During 1954–55, the station building was replaced with a new red brick building. This included a new ticket office, toilets and waiting rooms. Photo: Leslie Freeman © The Transport Treasury

October 1962 • Forres station looking west. Photo: Norris Forrest © The Transport Treasury

Final checks are carried out prior to the opening of the new Forres station. Photo: Ian Lamb

Class 158 No. 158706 at the head of a 4-Car DMU, the 09:00 from Inverness to Aberdeen, makes its presence known as it glides into Platform 1 with the first scheduled train over the new line. Photo: Ian Lamb